THE COMPLETE GUIDE TO GRASS-FED CATTLE

How to Raise Your Cattle on Natural Grass for Fun and Profit

By Jacob M. Bennett

The Complete Guide to Grass-fed Cattle: How to Raise Your Cattle on Natural Grass for Fun and Profit

1405 SW 6th Ave. • Ocala, Florida 34471 • 800-814-1132 • 352-622-1875–Fax
Web site: www.atlantic-pub.com • E-mail: sales@atlantic-pub.com
SAN Number: 268-1250

Library of Congress Cataloging-in-Publication Data

Bennett, Jacob M., 1979-
The complete guide to grass-fed cattle : how to raise your cattle on natural grass for fun and profit / by: Jacob M. Bennett.
p. cm.
Includes bibliographical references and index.
ISBN-13: 978-1-60138-380-8 (alk. paper)
ISBN-10: 1-60138-380-0 (alk. paper)
1. Cattle. 2. Cattle--Feeding and feeds. 3. Pasture animals. I. Title. II. Title: How to raise your cattle on natural grass for fun and profit.
SF197.B46 2011
636.208'45--dc22

2010042755

PROJECT MANAGER: Amy Moczynski
BOOK PRODUCTION DESIGN: T.L. Price • design@tlpricefreelance.com
PROOFING: Gretchen Pressley • phygem@gmail.com
COVER DESIGN: Meg Buchner • megadesn@mchsi.com
BACK COVER DESIGN: Jackie Miller • millerjackiej@gmail.com

Printed on Recycled Paper

Printed in the United States

We recently lost our beloved pet "Bear," who was not only our best and dearest friend but also the "Vice President of Sunshine" here at Atlantic Publishing. He did not receive a salary but worked tirelessly 24 hours a day to please his parents. Bear was a rescue dog that turned around and showered myself, my wife, Sherri, his grandparents Jean, Bob, and Nancy, and every person and animal he met (maybe not rabbits) with friendship and love. He made a lot of people smile every day.

We wanted you to know that a portion of the profits of this book will be donated to The Humane Society of the United States. ***–Douglas & Sherri Brown***

The human-animal bond is as old as human history. We cherish our animal companions for their unconditional affection and acceptance. We feel a thrill when we glimpse wild creatures in their natural habitat or in our own backyard.

Unfortunately, the human-animal bond has at times been weakened. Humans have exploited some animal species to the point of extinction.

The Humane Society of the United States makes a difference in the lives of animals here at home and worldwide. The HSUS is dedicated to creating a world where our relationship with animals is guided by compassion. We seek a truly humane society in which animals are respected for their intrinsic value and where the human-animal bond is strong.

Want to help animals? We have plenty of suggestions. Adopt a pet from a local shelter, join The Humane Society and be a part of our work to help companion animals and wildlife. You will be funding our educational, legislative, investigative and outreach projects in the U.S. and across the globe.

Or perhaps you'd like to make a memorial donation in honor of a pet, friend or relative? You can through our Kindred Spirits program. And if you'd like to contribute in a more structured way, our Planned Giving Office has suggestions about estate planning, annuities, and even gifts of stock that avoid capital gains taxes.

Maybe you have land that you would like to preserve as a lasting habitat for wildlife. Our Wildlife Land Trust can help you. Perhaps the land you want to share is a backyard — that's enough. Our Urban Wildlife Sanctuary Program will show you how to create a habitat for your wild neighbors.

So you see, it's easy to help animals. And The HSUS is here to help.

2100 L Street NW • Washington, DC 20037 • 202-452-1100

www.hsus.org

Author dedication

For my parents, for Gwen Mills, and for everyone who offered me a sandwich or a couch during the Great Recession.

Author acknowledgements

To write this book, I needed a ton of help from a lot of smart people.

Thanks to all the farmers, foodies, researchers, and professors who granted phone interviews, double-checked chapters, or invited me out to see how it's done. Carrie Balkcom of the American Grassfed Association answered more than a few phone calls and e-mails. So did Alan Yegerlehner, Joe Horner, Lee Meyer, and Greg Halich.

Hands down, the person I relied on the most was my dad, who was always there when I needed to know something about cow placenta or spontaneous hay combustion. My favorite part of this project was going over stuff like this at the picnic table on the farm. Thanks, Dad.

And thanks to Mom for reading to me in the first place.

There were people in my hometown, or people passing through, who helped me in my efforts to get my facts straight: Kyle Hardesty, Kristy Kibler, Eric Hawkins, and Andy Mills.

I also owe a shout out to anyone who ever opened an e-mail titled "Can you read this?" The list is long but distinguished: Ryan Clark, Taylor Loyal, Kate Corcoran, Erica Walsh, Rebecca Coudret, Joni Hoke, Ryan Reynolds, Linda Negro, Charlene Tolbert, Steve Arel, Jamesetta Walker, Rob Kaiser, Travis Bradford, Brandon Mattingly, and many others I'm sure I left out. Thanks, too, to all the English teachers and journalism professors who whipped me into shape in Meade County and at Western Kentucky University, and to every editor I've ever had.

A lot of people went out of their way to help me out, which made writing this book fun and informative. I hope you find reading it to be, too.

Table of Contents

Chapter 5: Building Your Herd 113

Introduction

When his stepdaughter's health depended on eating right, Rob White committed to raising grass-fed beef. On his Snelling, California, farm at the foothills of the Sierra Nevada Mountains, White used to raise cattle to sell at auctions to large-scale beef producers. He'd previously experimented with raising cattle on an all-grass diet when he was concerned about his high cholesterol, but he did not make the decision to produce only grass-fed beef until his stepdaughter, Summer Vitelli, was diagnosed with a rare blood disease after the birth of her first child and her life depended on eating the healthiest food available. As he and his stepdaughter researched ways to improve their diets, they discovered beef from hormone-free, antibiotic-free, grass-fed cattle is more nutritious than conventionally raised, grain-fed cattle. They became convinced raising their own grass-fed beef was the way to go.

"We got aware of what we were eating," said White, who now calls his beef business Foothill Grassfed Farm. "She had to be careful. She had to watch everything."

White is now in his third year of grass finishing his cattle, which means he raises his animals to the weight they need to be for slaughter completely on a diet of pasture and hay — and without the grain most beef farmers rely on to fatten their animals. The family owns about 60 head of cattle. His goal is to own about 100 breeding cows and to grass finish about 30 animals each year.

White makes most of his money by growing and selling almonds from the trees surrounding his pastures, but he also raises chicken and pigs for pork on his farm to further reap the benefits of organic, homegrown food. Since he made the switch to grass feeding, his cholesterol is down from 289 to 200. His level of triglycerides, a type of fat, dropped from 480 — "crazy high," he said — to 150.

After a long battle with her blood disease, Vitelli seems to be in remission. White, a retired English teacher, finds happiness with his new farming philosophy. Aside from experiencing health benefits, he gets to live the rural life of a cattle farmer while also creating a product that attracts Los Angeles transplants and Bay Area commuters who are concerned with animal rights and who take an interest in where their food comes from.

"Plus, I like steak," White said. "I did not like thinking that steak I was eating was killing me."

White is part of a small but growing movement of grass-fed cattle farmers. These farmers buck conventional agriculture practices and commit to giving their herds the most natural lifestyle possible. In addition to benefiting their animals, grass-fed cattle farmers improve the health of their fields by managing the herd's grazing patterns so the animals do not eat the same plants too often, which allows the pasture grasses to establish stronger root systems and allows them to retain more water and nutrients. Well-managed pastures also reduce the need for fertilizers, herbicides, and supplemental feeds. Pasture grasses are the most nutritious food you can feed cattle, so these animals produce the most nutritious meat and dairy products available on the market. This, in turn, leads to healthier consumers.

The grass-fed cattle movement is counter to conventional cattle farming, which is dominated by big machinery, big herds, and big farms. The biggest farms are called Concentrated Animal Feeding Operations, where cattle are kept in confinement and given growth hormones and antibiotics to prevent them from getting sick during the unnatural growth process. These cows are fattened using grain, not their natural food source. Raising the grain to feed these animals exacts a heavy toll on the environment. Millions of acres are covered with pesticides and synthetic fertilizers, and the grain is harvested using gas-guzzling machinery. Crop fields are also prone to erosion. The meat from these cows is often trucked thousands of miles around the country, which contributes to fuel emissions. It is a system of food production that critics, including small farmers and environmentalists, call unsustainable.

Consumers are increasingly aware of the benefits of buying from small farmers. The potential perils of industrial cattle farms came to the forefront of Americans' minds after recent outbreaks of E. coli bacteria and Bovine Spongiform Encephalopathy (BSE), better known as mad cow disease. In 1999, it was estimated that E. coli sickens about 73,000 people and kills 60 people in the United States each year, according to the Centers for Disease Control; it is believed those numbers have been dropping since then. Through February 2010, there have been 21 cases of BSE in North America, including three in the United States, according to the Centers for Disease Control and Prevention. Movies such as *Food Inc.* and books such as *The Ominivore's Dilemma* by Michael Pollan have criticized factory farms for their inhumane treatment of animals in unsanitary conditions, as well as the toll these farms take on the environment. A growing number of consumers want to know their beef and dairy products are raised humanely and without potentially harmful hormones and chemicals.

The market for grass-fed beef and dairy products is still a small niche, but it has huge potential. Grass-fed cattle farmers can sell their milk right off the farm, and farmers can mail their grass-finished beef all over the world, often to families who cannot find these types of products locally. It can sometimes be a challenge for farmers and customers to find each other, but cattle farmers are finding new ways to advertise their products all the time. There is room in the market for both farmers who want huge herds and others who want just a few animals. White said he uses a modest website, **www.foothillgrassfed.com**, to market his beef, but he said if interest grows much more, he will have trouble keeping up. He would rather keep his cattle business relatively small.

"The almonds pay the bills, but there is not a whole lot of personality in the almonds," he said. "The part I enjoy is being out with the cattle. I enjoy dealing with the cows and dealing with the people. I have met a lot of neat people. They want to know where their food comes from, and they want to see where it's grown. You cannot do that in a grocery store."

Farmers who once practiced conventional farming methods, such as using grain or hormones, find fulfillment in the switch to grass-fed cattle farming. It can be gratifying to get to know your customers and to know you are providing them a top-notch, nutritious product. Although they are careful to say it is not cheap or easy, some experienced grass-fed farmers also say newcomers may have an advantage over lifelong conventional farmers because they do not have bad habits to break. They also say grass-fed cattle farming is a good way for newcomers to begin farming because you can often get by with just the basics. You can get started with just a few acres of rental pasture, a handful of animals, some electric fencing, and portable water tanks. Starting a dairy farm requires more of an investment in facilities and equipment, but grazing-based dairies still can be profitable.

This book will show you what you need to know to start a grass-fed beef or dairy farm. It includes tips on finding farmland to fit your goals and discusses the steps to take to improve soil and plant health. You will learn how to choose the right grasses and the right animals so your operation can thrive in your environment. You will find tips on properly managing your herd to get the most out of your pasture, the cheapest and most nutritious way to feed cattle. You will learn how to time your cattle's life cycles to take advantage of the grass growth cycle and how the two benefit each other. This book includes information about basic cattle health care and how to identify and treat sick animals. This book also explains all the steps of functioning dairy and beef operations, including collecting milk, slaughtering beef animals, and marketing your products to appeal to potential customers.

A grass-fed cattle operation is a long-term commitment. It will take time to find the right mix of grasses in your pasture and the right animals for your herd. The success of your operation hinges on your own knowledge and observations, which will take time to develop. Minimize mistakes and optimize success with careful planning, hard work, and the effort to learn from others.

This book includes tips from grass-fed cattle farmers from around the country, from operations big and small, new and established. They offer this advice in the hopes you can learn from their successes — and from their mistakes. They offer their advice because they believe in their management practices and in the importance of sustainable agriculture. They are honest about the risks involved; the profit margins can be slim, and the opportunities for things to go wrong are plentiful. Each step in the process, from caring for land and animals, to marketing the finished products, is a challenge. But grass-fed cattle farmers are passionate advocates for this method of production. They believe in treating animals humanely, and in giving them the best foods so they can produce the best milk and meat. They believe this is the only way to farm for a sustainable future. And as these farmers will tell you, it can be a lot of fun.

0691 307

• Chapter 1 •

Why Cattle and Grass Need Each Other

TERMS TO KNOW

Grass-finish: The process of using an all-pasture diet to fatten a beef animal so it can be used for meat

Cud: Pieces of food regurgitated and chewed again

Ruminants: Even-toed, hooved animals that chew cud, including cattle, bison, goats, and sheep

Rotational grazing: The practice of dividing pasture into sections and rotating herds from one section to the next. It is also known by other names, including managed grazing and management intensive grazing.

Paddock: A subdivided section of pasture in a rotational grazing system

Forage: Food that grazing animals eat. Generally understood to mean grasses, legumes, or leaves. Forage can also be used as verb to describe the act of searching for and procuring food.

Customers who buy grass-fed beef and dairy products want something that goes beyond the standard way of doing things. Besides wanting tender, flavorful meat and delicious milk, these customers are looking for the most nutritious products on the market. They are looking for milk and meat free of hormones and antibiotics, and they want to know the animals were treated humanely and allowed to live naturally instead of being crowded on uncomfortable feedlots. These customers want to support family farms, especially those in their local communities.

What is Grass-fed Cattle Farming?

There is sometimes confusion about what grass-fed cattle farming is. True grass-fed cattle farming differs from conventional farming because the animals are left on pasture year-round, where they get most, if not all, of their nutrition from the plants in the field. True grass-fed beef cattle are **grass-finished,** which means they are grown to their market weight on a pasture-based diet, without grain. The all-grass diet improves the nutritional content of the resulting beef and dairy products. On conventional cattle farms, animals spend at least part of their lives penned up on feedlots or in dairy houses. These animals are also fed grain. Dairy cows are given grain to boost milk production, and beef cattle are given grain to fatten them up quickly during the last months of their lives. A grain-based diet is not natural for cattle and does not provide the same nutritional content as an all-pasture diet.

For the purposes of this book, "grass-fed" will refer to cattle that live their lives on pasture and eat only pasture and forage-based diets, with no grain. When this book speaks of grass-fed beef, it is assumed the animals were "grass-finished," with no grain.

The various labels used for beef and dairy products can be confusing to consumers. The U.S. Department of Agriculture (USDA) has a definition for products marketed as grass-fed: grass-fed cattle can only be fed mother's milk before weaning, and then, they must be fed only forage or roughage-based foods such as hay; they can never be fed grain. This definition further states animals must have continuous access to pasture during the growing season. But it is not a legal standard, and the USDA's definition for marketing products as grass-fed is voluntary for farmers, so only meat and milk that carry the "USDA Process Verified" label guarantee the product came from an animal never given grain. There are many beef and dairy products on the market identified as grass-fed, but not all of these products came from animals raised entirely on pasture. Some farmers label their beef as grass-fed because their animals spent most of their lives in pasture, even though their production practices do not match the USDA's definition — for example, they might keep the animals on pasture most of the year and still feed some grain. In such cases, farmers must explain on their label how these animals were raised on their farm. *Labeling will be discussed in more detail in Chapter 10.*

Many grass-fed cattle farmers believe the USDA's standards do not go far enough; for example, the USDA's grass-fed definition does not outlaw hormones or unnecessary antibiotics commonly given to conventionally raised beef. These farmers also believe the "access to pasture" part of the rule is too vague and leaves room for farmers to confine their animals unnecessarily.

American Grassfed Association definition

The most comprehensive standards for raising livestock on pasture come from the American Grassfed Association (AGA). The AGA's standards apply to all food produced from domestic **ruminants**, even-toed, hooved animals that chew cud. According to the AGA website, "The AGA defines grass-fed products from ruminants, including cattle, bison, goats, and sheep, as those food products from animals that have eaten nothing but their mother's milk and fresh grass or grass-type hay from birth to harvest all their lives." The

AGA has asked the USDA for strict guides on grass-fed marketing claims so products match a consumer's perception of what grass-fed products should be. The AGA felt the USDA's rules for raising grass-fed ruminant animals are not detailed enough and leave room for the misinterpretation of product labels by consumers who think they are purchasing food from farmers who used good animal husbandry practices, did not confine their animals to pens or feedlots, and did not use hormones or antibiotics. In 2008, the AGA began its own third-party grass-fed certification program. AGA-certified farmers undergo on-farm audits to ensure animals are cared for humanely, are allowed to live on open pastures, and are not given antibiotics or hormones. The goal is to produce nutritious and healthy foods and to support American family farms. *The AGA's guidelines for feeding and caring for animals are printed in Appendix B.* For the most up-to-date AGA information, visit its website at **www.americangrassfed.org**.

Differences between grass-fed and organic

Customers may not know there is a difference between organic and grass-fed cattle. Although cattle can be both organic and grass-fed, grass-fed cattle are not necessarily organic and organic cattle are not necessarily grass-fed. Cattle with the USDA's organic label are raised by standards under the National Organic Program. Accredited agents handle these certifications.

The organic label is intended to assure customers farmers did not use synthetic pesticides and fertilizers on their fields or hormones and antibiotics on their animals. The organic label also requires the animals to be kept in pasture during the growing season, defined as at least 120 days per year. They must receive 30 percent of their dry matter intake from pasture during the growing season. Dry matter is the feed in plant material after moisture is removed. The pasture requirement is waived for organic beef during the finishing period, when most beef farmers provide their animals a grain-based diet. There is still debate in the industry about whether this should be allowed for organic cattle. Although some grass-fed cattle are raised following the organic principles of using no hormones or synthetic substances, the organic label does not guarantee beef and dairy products came from grass-fed animals. These organic products could have come from cattle that were fed organic grain.

Many grass-fed cattle farmers, especially those who sell their products in large markets where they may not have direct contact with consumers, strive for the organic label. The organic certification exists as an assurance to customers that the products they are buying were raised to a certain standard. Small farmers who have relationships with their customers may not need the label as assurance because they can answer questions about their production standards and even invite customers out to have a look for themselves. Some grass-fed cattle farmers who sell their products directly to consumers adhere to organic principles, such as no synthetic chemicals or antibiotics, but do not seek the actual certification because there is a lot of paperwork and because it can be expensive to buy certified organic feed and fertilizers.

Grass-fed cattle obstacles

There are challenges to grass-fed cattle farming. Grass-finishing beef cattle often takes two years, sometimes more, whereas grain-finished cattle are usually ready in about 16 to 18 months. Another challenge to feeding cattle grasses year-round is the seasonal and geographic differences in grass availability. In many parts of the United States, heavy snow sometimes prevents grazing. In other parts of the country, dry conditions can hamper pasture quality.

Despite the various challenges presented in different parts of the country, farmers manage to grass-finish their animals in every region. **Graziers**, or farmers who manage grazing animals, use savvy management practices that mimic the way their animals' lives would work in nature to get the most out of their pastures.

Modern-day grass-fed cattle farming

In letting their cattle grow naturally on grass diets, today's grass-fed cattle farmers are returning to the farming techniques herdsmen used for centuries. But they are building on these centuries-old techniques by using current scientific data to make decisions that will improve the quality of their products. Farmers can test soil samples from their fields to show the soil's exact chemistry and nutrient content, which enables the farmers to make any necessary corrections so the most nutritious plants have the best possible environment to grow.

Science also helps farmers get the most out of each animal. Genetic data in categories about such things as growth rates and birth rates helps farmers choose the breeding animals most likely to meet their individual goals. This data is usually reported by farmers and maintained by breed associations.

There are dozens of breeds to choose from — some thrive in heat, and others are better adapted to cold. There is data available to show which breeds do well on all-grass diets and which breeds might not thrive in a particular region of the country. *The attributes of different breeds are examined in Chapter 5.*

The end result of all these processes is a product researchers have determined to be nutritionally dense. The best animals are eating the best grasses, which leads to optimum results and more consistent products. A farmer's beef and dairy products even develop a unique taste consumers come to expect and prefer, like a favorite wine from a familiar winery.

The Relationship Between Cattle and Grass

Cattle are perfect grass-eating machines. They evolved to break down the tough outer shells of grasses and legumes to extract the nutrients they need. When left to roam free in nature, cattle traveled in great herds, packed together to protect against predators. These herds grazed or trampled every plant in the way and deposited nutrient-rich feces and urine as they went. The cows' trampling loosened the soil for water absorption and prepared dead plants for breakdown. The herd continuously moved toward greener pastures, which left the just-eaten areas ready for re-growth from the nutrients and microorganisms in their waste.

Pasture grass is the most nutritious thing cattle can eat. Fresh pasture is a pure source of vitamins, minerals, and proteins grain or even harvested forages cannot match. The digestive systems of cattle enable them to extract these nutrients from tough grasses. Cows have four stomachs, and the first stomach, called the rumen, is basically a fermentation vat. On an all-grass diet, the pH in the rumen is a neutral 7, hospitable for bacteria that help break down grass. These organisms break down fiber in grass and turn it into useful nutrients and fatty acids for energy.

A grain diet disrupts the chemistry in the rumen by making it more acidic and allowing different bacteria to take over. The shift in chemistry is so dramatic that if the switch from grass to grain is not handled gradually, a cow will get sick and can even die as the increased acidity poisons its bloodstream. Another problem is cows produce lots of gas during digestion, usually released through belching. If the switch to grain is not handled carefully, the gas gets trapped in the rumen, and the cow swells up and suffocates. Even if the transition is handled carefully, a grain diet deprives cattle of the vital nutrients found in grasses. Grain-fed cows still are given hay, used for fiber, but their altered stomach chemistry does not allow them to extract any nutrients from it. Cows grow faster when fed grain, but the resulting meat will not contain nearly as many nutrients as it would if the cows had been allowed to grow on grass.

Health Benefits of Grass-Fed Milk and Meat

In 2009, *Forbes* magazine praised grass-fed beef as one of the healthiest foods on Earth. Beef and dairy products from grass-fed cattle are higher in nutritional content and reduce cholesterol, diabetes, high blood pressure, and cancer. You can actually see extra vitamins on the products. Although fat on grain-fed cattle beef is white, fat from grass-fed beef is yellow because of the stored vitamin A. Meat and milk from grass-fed cows taste different from the products made from their grain-fed counterparts. The beef has a stronger, more distinctive flavor some customers like right away, some do not like at all, and others must get used to — all depending on personal preference. Grass-fed milk is creamy at the top and has a sweeter taste some customers describe as "grassier." A handful of farmers prefer their grass-fed milk to be raw — not pasteurized — because they feel this process robs it of some proteins, enzymes, and beneficial bacteria. Many health officials say drinking raw milk is too risky because of the chance that other bacteria present could cause illnesses. *We will discuss this issue more in Chapter 9.*

Beef from grass-fed cattle is leaner than from grain-finished cattle, with less fat that equals fewer calories. According to Eatwild (**http://eatwild.com**), the website of journalist and grass-fed advocate Jo Robinson, a 6-ounce steak from a grass-finished steer contains 100 fewer calories than a steak from a grain-finished steer. Grass-fed meat and dairy products are higher in "good fats" and lower in fats linked to obesity, high blood pressure, and heart disease. Good fats include omega-3 fatty acids, which are linked to a reduced risk of depression, schizophrenia, and other mental illnesses. Grass-fed beef and dairy products also have three to five times more conjugated linolic acids (CLAs) than products from grain-fed cows. CLAs are shown to slow tumor growth and may reduce the risk of cancer. Food from grass-fed animals has shown to be higher in vitamin A, which boosts the immune system, and vitamin E, which reduces the risk of heart disease. Conversely, grass-fed products are lower in omega-6 fatty acids, which in high quantities are linked to obesity and heart disease.

The purity of the grass-fed, or pasture-based, method of raising cattle is attractive to many customers. Graziers point out raising cattle on pasture reduces the need for antibiotics because many factors associated with off-pasture confinement that cause illness — like standing on floors covered in animal waste — are eliminated. Grass-fed beef and dairy products are free from antibiotics and growth hormones used on factory farms. The Union of Concerned Scientists, a nonprofit group that advocates for a healthy environment, has estimated 70 percent of all antibiotics used in the United States are used for preventive purposes in cattle, swine, and poultry. This overuse of antibiotics in agriculture is blamed for increasingly common strains of antibiotic-resistant bacteria, some of which are potentially fatal, such as methicillin-resistant Staphylococcus aureus (MRSA). There are also concerns about residual levels of natural and synthetic growth hormones found in the meat and dairy products from factory farms and the effects these residues could have on a human's hormonal balances. These synthetic growth hormones may be associated with the earlier onset of puberty in girls, which is dangerous because earlier onset of puberty is associated with a possible greater risk of breast cancer.

Grass-fed cattle farmers note that recalls spurred by food-borne illness outbreaks come from animals harvested from industrial feedlots. Grain-fed cattle have stomachs with a higher acidity level, which can produce acid-resistant forms of E. coli. If these acid-resistant forms of E. coli spread to humans, the bacteria cannot be killed by the natural acid levels in a person's stomach. The acid in a human's stomach is better equipped to handle E. coli that come from grass-fed cattle.

An animal on an all-pasture diet is not a candidate to contract and spread mad cow disease. The disease, which can spread to humans who eat the meat of infected animals, is believed to spread when cattle are given feed containing the remains of other infected cattle, which can sometimes happen with feed mixes given in feedlots. It is extremely rare for even feedlot animals to eat the remains of other animals because this practice was banned in the United States in 1997, but a couple of feedlot cattle have tested positive in the United States since then. On a 100-percent forage diet, there is little, if any, chance your cattle will contract mad cow disease.

Improved Farm Efficiency

One of the benefits of a grass-fed cattle farm is improved farm efficiency. Feed costs are one of the biggest expenses in a cattle operation, and grazing is the cheapest possible way to feed an animal. Grass-fed cattle farmers use a clever grazing management practice called **rotational grazing**, or management intensive grazing, that involves using fences to divide pasture into sections called **paddocks**. Cattle are left in a specific paddock until they eat the grass down to a certain level, usually 3 inches or more off the ground. Then, they are moved to the next paddock and remain there until that paddock is grazed down. The time in each paddock can be as short as one day to as long as four or five days.

The goal of rotational grazing is to balance the nutritional needs of the animals with the nutritional quality and growth cycles of the plants. By the time cattle are rotated back to the first paddock, the plants will have had a chance to recover. As your grazing system allows your cattle to get more nutrition from the field, you will save money because you will not have to spend so much on feed. The longer you keep your cattle in the pasture, the less harvested feed you have to provide.

The alternative to rotational grazing is the common management choice called **continuous grazing,** where animals always are kept in the same pasture area. Cattle graze their favorite plants too low and too often, which weakens the plant's root systems. The less desirable plants are avoided, which gives them a chance to go to seed and multiply, leading to more undesirable plants. During continuous grazing, cattle manure is concentrated in high-traffic areas, such as in shade or near water, rather than distributed across the pasture where the nutrients would be more helpful. By moving animals frequently, this helps to disperse manure over larger areas. Cows produce dozens of pounds of manure per day. In confinement or in a continuous grazing system, this is a problem, but in a well-managed, rotationally grazed pasture, this is a benefit.

Grazing-based cattle operations can be profitable by offering a product consumers will pay a premium for. But they also survive by keeping inputs low. Recently, conventional cattle farmers also began to implement rotational grazing systems to augment their systems and reduce costs. Here is a list of ways grass-feeding cattle improves farm efficiency, compiled with input from interviews and the USDA report "Profitable Grazing-Based Dairy Systems":

- Pure pasture is the least expensive and most nutritious feed source you can provide cattle.
- A pasture-based operation requires less of the labor, machinery, and fuel associated with harvesting, transporting, and storing forage.
- Cows are healthier, and veterinary costs are reduced when left in pasture rather than confined in feedlots, enclosed areas where beef animals are kept to be fattened before slaughter.

- Grass-fed dairy cows have longer reproductive lives and fewer reproductive difficulties than their grain-fed counterparts.
- Fertilizer costs are reduced because cattle manure is deposited relatively evenly throughout the pasture. This distribution reduces high concentrations of nitrogen in high-traffic areas and cuts the chance of soil pollution through runoff in high-traffic areas. Grazing cattle return as much as 80 percent of key nutrients such as nitrogen, phosphorous, and potassium to the pasture where plants can access them to use in their own growth cycles.
- Grasses given a chance to rest between grazings develop strong root systems that enable them to better withstand summer heat. *Grazing strategies will be discussed further in Chapter 4.*
- Soil health also improves in grazing operations. Well-managed pastures contain more organic matter, decomposed plant or animal material in soil. Organic matter is important to pasture health because it holds nutrients that can be released into the soil and helps hold water in the soil. Well-managed soils also are easier for water to penetrate, which decreases erosion. The need for irrigation is reduced.

Aside from the list of practical reasons to raise cattle on pasture, many farmers have reasons beyond economics for grass-feeding their animals. Some get excited about the nutritional benefits their products provide; others are drawn to the positive environmental impact pasture-based farming can have. It may also be helpful to compare the pros and cons of grass-fed and conventional cattle systems:

Grass-fed systems: Allowed to eat grasses, hay, and other plants. The animals spend nearly their entire lives in pasture.

Pros:

- Grass is cheapest type of food you can give cattle, and grass production and quality increases under a well-managed grazing system.
- Milk and meat from grass-fed animals have the highest concentration of vitamins and minerals.
- Grass-fed beef and dairy products are usually sold in niche markets where discerning customers will pay a premium for the health benefits and the knowledge that the animals were treated humanely.

Cons:

- Grass is not always available year-round. In many areas of the country, you might have to extend the grazing season by planting different types of grasses or other plants, or rely on stored feed such as hay.
- Grass-fed animals are often smaller and take longer to finish, sometimes 24 months or more.
- Although grass-fed cows can produce milk in quantities comparable to those of grain-fed cows, some grass-fed dairies produce lower volumes than conventional dairies.

Conventional: Animals are fed pasture for the first few months of their lives before switching to high-grain diets. Conventional farmers can also use hormones or antibiotics. There are small conventional farms that operate similarly to grass-fed systems, but they do feed grain, can remove animals from pasture for prolonged periods, and can give antibiotics.

Pros:

- Grain-finishing is the quickest way to fatten a beef animal.

- Grain-finishing provides a consistent product that is not threatened by the unpredictability of nature and the weather.
- Most customers in the United States prefer grain-finished beef and are used to mass-produced milk.

Cons:

- Grain is not a natural diet for cattle. It can make the animals sick and kill them.
- Most conventional beef and dairy products are produced on large, factory farms that confine animals to small, uncomfortable spaces.
- Large confinement operations often lead to health problems for animals, which increases vet bills and often leads to short lives for the animals.
- Conventional farming often relies on synthetic chemicals, such as pesticides and fertilizers, that damage the environment.
- Meat and milk from conventionally raised animals are lower in nutritional content than from grass-fed cattle.

In the next chapter, you will learn about things you should consider before you start your own farm. Get tips on budgeting, finding good land, and preparing your pasture for use.

CASE STUDY: U.S. WELLNESS MEATS

John Wood
Chief executive officer
U.S. Wellness Meats
Monticello, Missouri
www.grasslandbeef.com
eathealthy@grasslandbeef.com

John Wood, a fifth-generation farmer from Monticello, Missouri, could not believe the results when he taste-tested the first cow he raised completely on pasture in 1997. "We could not believe it was tender, we could not believe it had a good flavor," Wood said. "We just completely blew it off as a fluke." He harvested another pasture-raised animal the next year, and again did not trust the results — it was also delicious and nutritious. When Wood harvested four or five pasture-raised animals in 1999, tests done at nearby universities confirmed the tasty meat was packed with nutritional content that cannot be found in grain-fed animals.

"I made the fateful decision to bail out of the conventional family business and start something brand new," Wood said. "I like the unknown; I like a challenge, and so I have kind of thrived in this wild and woolly environment of grass-fed beef."

Soon after he completed his farm's conversion to high-quality pasture, Wood's family partnered with five other families to found U.S. Wellness Meats in 2000. U.S. Wellness Meats is part of a family farm cooperative that sells products, including grass-fed beef, grass-fed lamb, grass-fed bison, compassionately raised pork (pigs that are outdoors every day, never allowed on slatted floors, have fresh bedding available at all times, and live on a complete vegetarian diet), wild seafood, and rabbit cuts. The company ships these products all over the United States and to Puerto Rico, Bermuda, Hong Kong, and Iraq. Their specialty product is an energy bar from a Cherokee Indian recipe involving pemmican, a mixture of jerky and fat.

Wood first decided to experiment with holistic land management practices, which means managing land and animals in ways that mimic nature, in 1993 after attending a lecture on the subject. He liked the idea of using sustainable farming methods to raise healthier pastures. He signed up for the federal Conservation Reserve Program to raise a field of Eastern Gamagrass, a native, warm-season grass that covered the prairies in Missouri in the 1700s. Three years later, he opted out of the CRP contract early so he could use the field for seed and pasture.

His company's success was never assured. Wood said he and his partners learned everything about the business the hard way. They lost two partners in the beginning, and the others toughed it out over a two-year period in which the business did not profit. Their profit margins today are still slim. The partners thought the Internet would provide an instant market for them, but in the first two months, they only had about 35 sales, and only one of those orders came from someone they did not know.

"We realized how hard it was going to be," Wood said. "It was like trying to climb up a glass wall. Nobody knew what grass-fed beef was. You do not have any money to advertise, so you start going to trade shows and approaching doctors, and you start just working your networking the best you can do. You do not have an e-mail list, you do not have a database, and you cannot reach people. It was just like roping in the dark."

Like many farmers who raise their cattle on pasture, Wood remained passionate about his decision because it was a more satisfying way to farm than conventional beef farming and because he knew he was creating a superior product. He said it is tough to get people to change their eating habits, but he continues to preach about the environmental, health, and social benefits of grass-fed beef.

"It is good for the land, it is a good life for the cattle, and it is good for the rural community because it sustains the rural community," Wood said. "You are catching sunlight with green plants, the animals eat the green plants, and they convert it to the appropriate fatty acid chemistry, so when you eat it as a consumer, you are getting a gift out of the land."

• Chapter 2 •
Getting Started

TERMS TO KNOW:

Replacement heifers: Females used for breeding. On dairies, they are also used for milking.

Stocker cattle: Cattle in the stage between weaning and the finishing stage. They typically weigh between 400 and 600 pounds.

Steer: A castrated male calf that was unsuitable for use as a bull

Animal Unit Equivalent (AUE): A measurement defined as a 1,000-pound animal, the average weight of a mature beef cow

Stocking rate: The number of AUE per acre a pasture can support

Nurse cow: A cow used to nurse her calf and the calves of other dairy cows

Cooperative extension: A nationwide educational service staffed by experts called extension agents who provide information to farmers, children, small business owners, and others in rural and urban communities

Extension agent: A person who provides information to farmers and others in rural and urban communities through cooperative extension offices

Feedlots: Areas where cattle are gathered to be fattened on a high-grain diet before slaughter

Backgrounders: A type of beef-cattle operation in which farmers buy weaned calves and fatten them up cheaply before selling them to feedlot operations.

To succeed as a grass-fed cattle farmer, you need a good plan. You have to figure out where you want to start, where you want to be a few years down the road, and how you want to get there. Decide if you want to raise cattle for beef, or dairy, or a combination of both. You will need land, animals that fit your system, and an idea of who might buy your products if you plan to sell. By thinking through your goals, you will be able to plan for your future needs and avoid problems down the road.

If you are switching from conventional methods to a grass-fed approach, you will face other tough decisions — you will have to make changes in your herd's lifestyle and the way you care for them. You may find your land cannot support your herd at its current numbers, or that animals in your herd are not cut out for the pasture-based lifestyle. You will have to find new markets if you decide to sell your cattle's products, and there could be a disruption in your income during the transition phase if you are currently raising cattle. Getting your land healthy enough to support cattle year-round and breeding the right cattle for your system will take time. But each year, more and more farmers commit to grass-fed cattle farming procedures, and many say they are happier for it. For first-time farmers and for cattlemen thinking of switching from conventional methods, it helps to think about why you want to be a grass-fed cattle farmer.

What Are Your Reasons for Farming?

People usually get into farming one of two ways: Either they are born into it or they decide to try it because they find the lifestyle appealing. If you are thinking about starting a grass-fed farm, what are your reasons for this? Do you enjoy the country life? Do you like working with animals and the land toward a set of clear goals? Would you mind doing whatever is necessary for the health of your animals, no matter the weather or whatever else is going on in your life? Do you have the patience to negotiate the learning curve as you gradually improve your farm? If the answer to these questions is yes, you may enjoy raising grass-fed cattle.

There are ways to try out the farming lifestyle without getting in over your head. Find a job or an apprenticeship on someone else's farm, enroll in training courses at nearby universities, or find a mentor whose farming methods appeal to you. One place you may find training options is a cooperative extension office, part of a nationwide educational service staffed by agents affiliated with universities. These agents provide information about farming in most rural areas. The more hands-on experience you get before you start your own farm, the more likely your new venture will succeed. This experience will also be valuable later if you seek financing for a new farm or farm improvements because lenders want to know you have the expertise necessary to succeed.

Who are you growing for?

Some people are satisfied growing just enough meat or milk to provide for themselves and their families. They enjoy the process of working with nature and like the idea of knowing where their food comes from. If this is you, then you will not need much to get your operation going: a few acres, a small number of animals, fencing, and some key herd-management equipment.

Most farmers want to sell to others for profit. Many depend on farm income for their livelihoods or for supplemental income. Depending on the type of animal you buy and its age and condition at the time of sale, purchasing an animal for your herd will cost at least a few hundred dollars, if not as much as a couple thousand dollars. It can take a couple of years to ready a beef animal for slaughter or for a cow to produce enough milk to pay for your investment in her.

If you want to make money, you need to have an idea of who you are going to sell to. Decide on your target market, and develop a plan for getting your product in their hands. Do you know where to make the connections necessary to sell directly to customers? Are there farmers markets or other groups in place to help farmers find customers? Is there a restaurant or grocery in town that might buy your products?

What kind of farm will you have?

The goals of all cattle farms are not the same. Some beef farmers are proud to say their animals are born on their farm and live there until time for slaughter. Other farmers focus on specific phases of cattle development and buy or sell animals at points between birth and death. For example, stocker-cattle farmers handle animals in the phase between weaning and the finishing stage. Some dairy farmers will hire contractors to raise their heifers — females that have not given birth — so they can concentrate on their lactating animals. All of these systems have their merits. Here are some cattle-raising strategies you should be aware of:

- **Cow-calf**: Cow-calf farmers raise cows and bulls with the goal of producing calves. These calves are sold later to other farmers who will continue raising them for beef, sold to farmers who will use them as breeding stock, or kept on the farm until they are ready to be slaughtered for beef. If you wanted to start a cow-calf operation, you could purchase pregnant cows or cows with young calves. Calves can be sold after weaning when they are a few months old. Some farmers sell all their calves at weaning. Others keep them until they are yearlings, which means they are between 1 and 2 years old and have had time to put on more weight. This can increase the amount you make on each animal but only if you have enough grasses to feed them cheaply during winter. There are also farmers who breed purebred animals and sell the calves as breeding stock.

- **Stocker cattle**: Some farmers concentrate on stocker cattle, beef cattle between the weaning and finishing stages when they add an extra 200 to 400 pounds to reach their mature weight so they are ready for slaughter. Finished weight depends on breed; most grass-fed cattle breeds are finished at about 1,000 to 1,200 pounds. Stocker cattle typically weigh between 400 and 600 pounds when they are purchased after weaning. These farmers buy calves from cow-calf farms. The calves can be grown on pasture and then sold as feeder cattle, cattle ready for the finishing phase. Stocker operations have more flexibility than cow-calf operations because they do not have to wait out their cattle's natural breeding and nursing cycles to buy or sell animals. They can buy and sell animals at different times of the year, including selling their entire herd in the fall so they do not have the expense of winter feeding breeding herds must deal with.

- **Grass finishing**: You can grass finish calves born on your farm or bought from someone else. Grass-finished beef is a niche product that provides extra value per pound for your meat because customers appreciate the health benefits of animals raised in more natural environments than those of conventional beef. The main challenge of grass finishing is to keep animals gaining weight their whole lives with no periods of loss. This can be especially tough when grass stops growing in winter. It also usually takes longer to grass finish animals than grain-fed cattle.

- **Farmstead dairy processing**: Although many dairy farmers simply sell their milk to other companies, many grass-fed farmers find that they can make more money by processing their milk into other foods and selling packages of their own brand. You can use milk to produce cheese, yogurt, or other products, and sell them in stores on your farm or other local outlets. Dairy beef — dairy steer calves, castrated male calves — can be raised on your dairy farm or bought cheaply from other dairy farmers and raised as beef for extra income.
- **Dairy replacement heifers**: Dairy farmers who want to focus on their lactating animals sometimes contract with someone else to raise their replacement heifers. Heifers are young females that have not given birth. Replacement heifers are females that will be used for breeding, and on dairies, they will be milked after giving birth. Farmers use replacement heifers to step in when they get rid of cows that are no longer suitable for breeding, such as a cow that is too old to breed anymore. Sometimes these heifers are raised by a contractor who will return them to the owner shortly before birth. Some cow-calf farmers also make money by selling replacement heifers to other farmers. Caring for replacement heifers might not be a job for a beginning farmer because the other farmers might not trust a novice to take care of their valuable heifers.

Also consider if you want to be certified by a particular organization because you will have to prove to certifiers that farm decisions, big and small, comply with these standards. Many certifying agencies, including the AGA and the National Organic Program (NOP), require documentation, such as receipts or tags from medicines and feed supplements you use on your farm. The AGA designed its record-keeping system not to require much time.

Farmers who choose to be certified, either under the NOP, by the AGA, or by other organizations, do so because these third-party certifications serve as assurances to customers that the farmers adhered to certain standards. Many customers will pay more for these products; for example, customers who buy from AGA-certified farmers want products from animals that were allowed to live as naturally as possible — outdoors, eating grass. Customers who buy organic-certified products want to know their foods were created without the use of potentially harmful synthetic chemicals.

Under the NOP, a field is not considered organic until three years after the last use of synthetic chemicals. Still, many farmers do not mind the additional challenges of finding organic fertilizers, pesticides, and feeds because they believe it will pay off in the long run and because they believe it is the right thing to do. Learn more about organic certification at the NOP website at **www.ams.usda.gov/NOP**. One note of caution: It takes considerable skill to run a completely organic farm. If you are a first-time farmer, it is probably better to concentrate on building your farm management skills, such as keeping your animals healthy naturally and building your soil and plant health before seeking organic certification.

To be certified under the NOP, you must keep records of the substances you use on your farm and submit to inspections from accredited certifying agencies. At the beginning of the process, you have to fill out a farm plan detailing these steps and carefully examine each step of your production process. Everything, including seeds, fertilizers, and pesticides, has to be organic. You will have to work with certified organic processors, which takes extra time and commitment. Forty percent of organic dairy producers say the most challenging aspect of organic milk production is the certification paperwork and compliance costs, according to the USDA.

To actually use the word "organic" on the label, you must be certified under the NOP. However, other certifying agencies strive for principles similar to those of the NOP but aim to be simpler so small farmers are able to achieve these standards. One example is Certified Naturally Grown (**www.naturallygrown.org**), which seeks to reduce much of the government paperwork that makes getting certified under the NOP prohibitive.

Determining your ideal herd size

When first establishing your herd, it is better to start out small and grow into an operation than to start big and get in over your head. Two important things for a first-time cattle farmer to learn are how to handle animals and the skills of rotational grazing to improve your pasture. Some people only want to raise enough animals to provide milk or meat for their families. These farmers only need a minimal number of animals. For milk, you might only need a cow or two. For beef, you might be able to buy two or three weaned calves every year.

If you want a few cows, but do not want the expense of a bull, start with weaned calves that you raise as stocker cattle. Let them grow larger on your pasture, then sell them to another buyer who will finish them, or grass finish them yourself. If you start your cattle operation with recently impregnated cows, you will not need a bull for another year. If you start with new cow-calf pairs and you plan to re-breed the cows the next year, you will need a bull within about three months. You do not need a bull if you plan to use artificial insemination. *Artificial insemination will be discussed in more detail in Chapter 5.*

If you are serious about starting a cow-calf beef operation, use one bull for every 20 to 30 cows as a rule of thumb — if you have the land and skills to support it. If you feel that might be too much to handle at first, you could start with ten cows and perhaps share a bull with a neighbor. For example, if you plan to calve in the spring, you could share a bull with someone who plans to calve in the fall. Plan on keeping two to four calves for your beef business and sell the rest at sales barns, where they are usually bought by large producers as part of the commodities market. Large numbers of buyers and sellers congregate at sale barns for auctions. Buyers usually include feedlot operators who finish their purchases for beef, or backgrounders who fatten up the calves more before selling them again.

If you start small, you may have more land than your cows can keep up with, so just give them a section of your pasture you can subdivide into smaller sections. Then, rotate your herd through these subdivided sections. You could take the sections of pasture you do not use for grazing and cut them for hay, which you could sell or use for your own reserves.

If you want a larger operation, your herd size will ultimately depend on your land, as your pasture can only support a certain number of animals. Knowing your financial goals will give you an idea of what you want your farm to look like in a few years. If you already own a farm, a big part of the equation is filled in for you: Your maximum herd size will be determined by how many animals your land can support. The term "animal unit equivalent" (AUE) comes into play here. This measurement is calculated based on the nutritional requirements of an animal relative to a 1,000 pound cow with or without a nursing calf. The calculation assumes that 1 animal unit will consume 26 pounds of dry matter per day. Dry matter is the feed in plant material after moisture is removed. As you can see from the following table, not every animal will count the same, and the same animal will be different at various points in its life. A calf is worth .50 of an animal unit, but when it grows to 1,000 pounds, it will be worth a whole animal unit. A mature bull is worth more than 1 animal unit.

Animal	Weight (in pounds)	Animal Unit Equivalent (AUE)
Young cattle	500	.5
Cow	1,000	1.0
Bull	<2,000	1.5

Stocking rates are the number of AUE per acre you keep on your pasture. **Carrying capacity** is the stocking rate your pasture can support. The easiest way to figure these is to ask a local farming expert such as a county extension agent who can give you an idea of what pastures in your area similar to yours can carry. The average available dry matter per year depends on the area of the country and the quality of the pasture. Not every acre of your farm will be good pasture; for example, heavily wooded areas do not produce much grazing grass. The amount of dry matter available per acre per year can be as little as 2 tons to more than 5 tons. So if your field yields 3 tons per acre per year and you have a 50-acre farm, that is 150 tons per acre, which equals 300,000 pounds per acre. Deciding how many cows to keep is an inexact science. Varying weather conditions from year to year will cause fluctuations in your pasture growth. But if you want to read more about making your own estimates, try "Forage Production and Carrying Capacity" from the University of Idaho (**www.cnr.uidaho.edu/what-is-range/Curriculum/MOD3/Stocking-rate-guidelines.pdf**) or "Stocking Rate: The Key to Successful Livestock Production" from Oklahoma State University (**http://pods.dasnr.okstate.edu/docushare/dsweb/Get/Document-2050/PSS-2871web.pdf**).

A formula you can use to figure stocking rate is the estimated amount of available dry matter in your field per year divided by the amount of dry matter a cow eats a year, or

(Pounds of dry matter per year) ÷ (pounds of dry matter per year per cow)
= stocking rate

So if each cow needs 26 pounds of dry matter per day, that is 780 pounds per month, or 9,490 pounds per year. If your field yields an estimated 300,000 pounds per year, and each head of cattle needs 9,490 pounds per year, your field could support 31.6 head of cattle.

It is probably a good idea to estimate a bit low on stocking rate to provide a cushion for when your pasture growth is below average. In times when growth is above average, buy extra stocker cattle, or keep a few cattle you had planned to sell.

As you get better at managing your herd's rotation patterns, you will be able to provide more nutrition from grazing and rely less on stored forage and supplements. You may also be able to increase your stocking rates because your pasture yield will improve.

The Differences in Raising Cattle for Milk and For Meat

Dairy farms and beef farms can both succeed as grass-fed systems, but each type of operation poses its own challenges. Beef production and dairy production depend on different breeds of animals. You can buy beef and dairy animals from similar sources, such as public auctions or private sellers, but base your purchases on different sets of criteria. Different breeds are often used for beef rather than for dairy, and each type of operation provides differing management challenges.

With grass-fed cows you intend to raise for beef, take the following points into consideration:

- Your goal is to get the animals to fatten to a weight that will produce the best meat before slaughter. Your success depends on timing the animals' nutritional needs to the seasonal yields of your land. Grass-fed beef cattle may need 20 months or even two years or more to reach finishing weight, compared to a typical 16 to 18 months for grain-finished cattle.

- Beef calves are left with their mothers for between six and nine months. They can then be sold to other farmers or kept on the farm to continue to grow.

- Beef animals are usually killed at slaughterhouses and then cut up by butchers. Depending on who you plan to sell your beef to and the laws in your state, the facilities and the carcasses may need to be inspected by state or federal officials. *This will be discussed further in Chapter 9.*

If you plan on raising grass-fed cattle for dairy items, here are some points you should consider:

- Your goal is to get them to produce as much high-quality milk as possible. Dairy cattle are usually milked twice a day, every day — once in the morning and once in the evening — for much of the year. Many grass-fed dairies are seasonal, meaning they breed their cows at the same time and have a dry period where they do not milk cows, usually during the last two months before cows calve again. At this stage of pregnancy, cows use less of the nutrients they take in for milk production and put them toward their developing fetuses.

- Dairies depend on new mothers to provide milk to sell. On conventional dairies, calves are fed milk-replacer solutions. Because grass-fed dairies aim to give cattle as natural a feed as possible, these dairies must find other ways to provide milk for their calves. These methods include letting mothers nurse their own calves, diverting milk from the herd into barrel feeders for the calves, or using **nurse cows** that nurse their own calf and the calves of others. *The pros and cons of each method will be discussed in more detail in Chapter 8.*

- Dairy farms also require extra equipment such as milking machines and coolers.

- Each state has different laws governing milk sales. It is important that you speak to state agricultural officials to familiarize yourself with the laws in your state. You will probably need a license to sell milk to processors or milk companies. If you sell milk, government officials will have to inspect your dairy facilities.
- It is especially important to know your state's laws about selling raw milk, one of the most hotly debated issues in agriculture. Eleven states allow licensed dairies to sell milk to retail outlets, according to the National Conference of State Legislatures. Twenty states allow people to buy raw milk off the farm or get milk if they buy ownership shares of an animal. *We will talk more about raw milk in Chapters 9 and 10.*

What Will Your Budget Be?

You must develop a budget with realistic expectations. Investing in cattle, land, and farm equipment can be costly even if done frugally. You will have to account for variable operating expenses, such as gasoline and fertilizer. Unless you have family equity or other partners who can help cover these expenses, you will need to sell some of your animals or the products they produce on your farm to cover the expenses your new business venture generates.

Even good farmers do not last in the business if something goes wrong with their budget. A common mistake is investing more in capital than your animals can cover, said Joe Horner, an extension associate of agricultural economics for the University of Missouri College of Agriculture, Food, and Natural Resources. In other words, the farmer spends too much on land or equipment and does not have enough money left over to buy the animals needed to provide income for the farm. Horner said many young dairy farmers get into trouble because they check out the parlors of more experienced farmers, and the veterans tell them they wish they had built bigger parlors when they started out. So the young farmers go out and build parlors much larger than they originally planned, and then they cannot make their loan payments. Before the young farmers can grow to the scale they wanted, they get discouraged or go out of business. "What a lot of older people forget is that if they would have built to the scale they wanted to be operating at, they would have been broke before they ever got started," Horner said. "You have to start with what your cows can pay for."

Horner recommends purchasing cows before you purchase land. You can do this by leasing someone else's farmland. You could also lease an old dairy and upgrade its milking equipment so you can milk more cows per hour. The USDA Farm Service Agency (**www.fsa.usda.gov**) and many state governments have Beginning Farmer Loan Programs, and 70 percent of the funds available are set side for beginner farmers. Once that loan is paid off, you might be ready to purchase land. Horner said he has seen farmers in their 20s start out on a lease farm with a few cows, and by the time they are in their 40s, they are milking 500 or more cows.

It would be a good idea to find a financial professional to help with your budget planning. The U.S. Small Business Administration (**www.sba.gov**) and county extension offices have checklists and guidelines for creating business plans. There are loans available to purchase land, livestock, equipment, and supplies, or to make farm improvements. The U.S. Farm Service Agency (FSA) offers farm ownership or operating loans to family-size farmers who cannot get credit from banks or other institutions. These loans are available for beginning farmers who cannot obtain farming elsewhere and for established farmers who have experienced a difficulty such as a natural disaster or who have limited resources available to them. Find more information about these loans on the FSA site at **www.fsa.usda.gov**, or by visiting your local USDA Service Center. Search by state at their website, **http://offices.sc.egov.usda.gov/locator/app**.

The FSA also works with vendors who offer training in record keeping, planning, and goal setting. The Oklahoma State University Department of Agricultural Economics (**http://agecon.okstate.edu/Quicken**) sells add-on software to use with the budgeting software Quicken to keep track of farm expenses and income. Market fluctuations will affect the profitability of your business, and other costs, such as land or equipment, vary by area. The University of Kentucky College of Agriculture has budgeting examples and decision aid tools at **www.ca.uky.edu/agecon/index.php?p=565**.

Horner and colleagues at the University of Missouri created models for starting dairies from scratch, as well as for converting an old dairy. These models can be viewed at **http://agebb.missouri.edu/dairy/dairylinks/models.htm**. The professors figure it costs $1 million to build a 200-cow dairy from scratch, with an average cost being $5,375 per cow, and most banks will not lend more than half the amount to start an operation. This is usually a deal breaker because many people do not have $500,000 of equity. This discourages many farmers from trying to start their own dairy farms, but Horner said you can make it if you start small and do not take on too much debt.

"You have to start with the first step of the ladder," Horner said. "You may not be able to build that 200-cow dairy from scratch. You may be able to build 200 cows on somebody else's dairy, and then go build a 200-cow dairy."

There are other ways to control costs. Try to figure out what you absolutely need to own, what you can get by with renting, and what services you can contract out. For example, many farmers own hay-making equipment; others hire someone to cut and bale it for them or just simply buy hay from someone who is selling it. Many farmers own seeding equipment, but often you can rent these and not have to buy them. County extension agents or local cattle associations can tell you where to find rental equipment in your area. If you must purchase equipment, try to find it used, or try to find less expensive options. For example, instead of buying a tractor, you might be able to rent or buy a **skid steer**, a useful vehicle that can do much of the heavy lifting required on a farm.

Finding Good Land

Unless you already own some acreage or are taking over your family's farm, you will have to find land to farm. If you own land already, check with your local zoning office to be sure it is zoned for agricultural use and to see if there are usage or other restrictions. For example, land zoned for agricultural use could have limits on the number of animals you can keep on that property.

If you do not own land, you face another choice — should you buy or lease? The answer depends on your individual finances and where you live. For example, it is cheaper to buy farmland in Missouri than in California. If your finances are limited, you could save money in the beginning by renting land rather than buying it. If you start renting land, you will have more money to invest in the animals that will provide income to your farm. Once you have enough money coming in from your animals, then you could consider buying land. If you find good rental land at a reasonable price, you might be able to sign a long-term lease, usually three to five years, with the landlord. It is common for long-term leases to have termination clauses in case the landlord sells the property. Some farmers own or lease more than one site. Many farmers have homes on their farms, while others have a residence separate from their pastureland.

How much land do you need?

The amount of land you need depends on what your goals are. Small-scale farmers often only need tracts of land 40 acres or smaller. This is linked to the size of the herd you would like to have. Talk with a local farming expert about your goals to determine how much land you need. If you have to start with a smaller herd than what you want or what your land can support, rotationally graze those animals in a small number of paddocks and harvest the unused grasses for hay. If you lease land and keep your infrastructure investment small, you will be able to move to another site later if you decide you want a larger operation.

Where to find land

If you are looking for a site to farm, there are resources available to help you make your choice. A real estate agent should be able to help you scout your target area. There are "land-link" programs that connect young farmers with retiring ones based on each farmer's goal. These arrangements allow the young farmers to lease land from experienced farm owners or to partner with farm owners, often with the option to buy the farm after a predetermined period. Land-link programs include the International Farm Transition Network (**www.farmtransition.org**). The Beginning Farmers website (**http://beginningfarmers.org**) has many resources for new farmers, including a list of land-link programs. The Farm Service Agency advertises properties for sale and gives beginning farmers first priority at **www.resales.usda.gov**. The website allows you to search by state.

Qualities of a good site

When looking for your farm, consider many of the same factors you look at for any other real estate transaction. The old adage proves true: real estate is all about location, location, location. If you are going to live on your farm, it probably needs to be in reasonable proximity to all the important things in your life: school, family, jobs, and any other locations you need to travel to frequently. If you are going to lease the farm and live elsewhere, keep in mind you will need to be there every day, so it should be within a reasonable driving distance from your home. Other qualities to keep in mind as you search include:

- The ideal farm will have good access to roads. You are going to have to get trailers and other machinery in and out of there, so make sure there are no barriers to transportation.

- Trees can form a natural shade, but if there are too many or if they are packed too densely, you will have to clear them and plant edible grasses.

- You are going to need power to run electricity for water, fences, and tools used for cattle health care. A site with access to power is preferable; if you do not have access to electricity, you will need to invest in a battery, possibly a solar-powered one. The battery will power your fence charger, part of the fence that sends the electrical pulse through your fence. You could use a generator to power the fence charger, but this would be costly because of high fuel prices needed to run the generator.

- You might want to look for land with adjacent rental properties that could provide room for future growth.

You can tell some of what you need to know about a property by looking at it. Fields used for row crops or hay will take years to reach their potential as pastures, and they may not be suitable for grazing for several months. If the field was left to grow wild, you may have to clear stubborn weeds and undesirable growth before planting more desirable plants. This does not mean you should pass on these properties, but these qualities should factor into your decision.

Poisonous plants

Before you put animals onto a pasture, it is important to know what poisonous plants grow in your area. Some plants are poisonous at certain points in their growth cycles or in certain parts of their anatomy, such as in stems or in seeds. Johnson grass, for instance, contains cyanide that becomes available in the leaves after frost. Animals that eat poisonous plants will show symptoms, such as nervousness or low appetite, depending on what they ate. Call a veterinarian if you suspect plant poisoning. Your area extension agents can help you identify the dangerous plants in your field and advise on how to protect your animals.

Soil Tests

Before you get locked into a loan or a lease, get the soil tested to determine its pH level and nutrient content, which will determine how productive the soil will be. The test results will come with recommendations to improve your soil chemistry and pasture grass-growing capability. Soil testing is easy, inexpensive, and lets you know more about your soil and what to expect during the first few years you own your farm. Getting soil healthy takes time, and it is usually beneficial to test your soil to ensure it is usable for your farm. Soil samples can be taken to county extension offices, and these extension offices send the samples to testing laboratories at universities. These offices also have sample test bags or boxes, the forms you need to submit with your samples, and more information about soil testing. Commercial laboratories also test samples for you, though they usually cost more than university laboratories.

Although testing can be done year-round, the best times to test your soil are fall and spring because this gives you an idea of what chemistry and nutrient conditions will be like during peak growth. Avoid testing when the ground is too wet, too dry, or frozen because the soil conditions at these points will be different than typical soil conditions.

How to take a soil sample

Follow these steps when taking your soil sample:

- To gather a soil sample, you need a clean bucket and a shovel, spade, hand trowel, or a soil probe, a hand-held tool that removes a small soil core. Make sure your tools and your bucket are clean, but only use water to clean them. Also, do not use brass, bronze, galvanized tools, or rubber buckets because they contain copper and zinc that could contaminate the samples.
- If you have a large pasture, take more than one sample to accurately represent each area of the field. Each sample should represent an area that generally covers no more than 20 acres. An area of field used to grow a crop will produce different results than an area used for grazing. A sample from a pasture area will be made of cores or samples of dirt equal to roughly ¼ cup, which you remove from beneath the surface. The cores are mixed together to form each sample.
- If there is a problem area in the pasture, such as an area where grasses do not grow well, take a sample from there and keep it separate from the other samples.
- Avoid taking samples from areas where nutrient content may be skewed, such as areas used for manure or hay storage.
- Be sure to dig to the appropriate depth. In areas that are not tilled, take soil from a depth of 4 to 6 inches. To take a core, simply take your tool and extract enough dirt to fill about ¼ cup. The cores taken for each sample will add up to about 1 pint.
- Place all cores for a sample in the bucket. Use a different bucket for each sample, or collect samples one at a time. Use your digging tool to

break up the clods and mix the individual samples thoroughly. Spread each sample out in uncontaminated open spaces to air dry.

- After a sample is dry, place it in the sample bag or box you got from your extension office. Be sure to clearly label the samples from each area so you know which area of the pasture it comes from.
- Fill out an information form to help the test evaluator make recommendations for your soil. This information form specifies how you plan to use the sample area.
- Take the sample to your county extension office. Test results usually come within a couple of weeks.

What the tests show

Soil tests give you several measures of soil health. Your test provides recommendations on steps you can take to reach the ideal pH for your field. Generally, pasture soil pH should be between 6 and 7 because if soil is within this range, nutrients are readily available, and beneficial soil organisms are most active. The three main elements plants need are nitrogen, phosphorous, and potassium, commonly referred to by their initials NPK — N is for nitrogen, P is for phosphorus, and K is for potassium. In smaller levels, plants also need calcium, magnesium, sulfur, boron, copper, iron, chloride, manganese, and zinc.

Based on the results from your soil test, the report will provide specific instructions on how to boost your soil health. The recommendations usually consist of soil amendments, or concentrated nutrients such as fertilizer that you should add to your field. You will be told when and how much of each deficient nutrient to apply. It is important to follow these recommendations closely; guessing the amounts of amendments to apply will shift the soil chemistry but not necessarily improve it.

Fixing Your Soil

You must have healthy soil and plants to maximize the yields of your beef and dairy cattle. Many grass-fed cattle farmers consider themselves grass farmers, with one of the benefits of taking care of their grass being that it produces good beef or dairy products. The natural relationship between plants and your cattle will also help improve the soil.

Getting your soil healthy takes time, but it is important.

Follow the recommendations of your soil test. You can purchase both synthetic and organic fertilizers in pre-mixed bags that identify their ratios of nitrogen-phosphorous-potassium in a numerical series such as 10-10-10. A 10-10-10 bag of fertilizer is 10 percent nitrogen, 10 percent phosphorous, and 10 percent potassium. In this example, a 100-pound bag of 10-10-10 fertilizer would be 10 pounds of nitrogen, 10 pounds of phosphorous, 10 pounds of potassium. The other 70 pounds would be filler material to make the product easier to spread. Many different ratio mixes are available; one such mix is called urea (46-0-0, a source of nitrogen); another is called potassium sulfate (0-0-50, for potassium). You can also have custom mixes made at farm supply stores to match the specifications in your recommendation.

Another common soil amendment is lime. Lime is made of limestone or chalk and is used to reduce an acidic soil pH. If it is recommended that you add lime, it should be applied several months before planting because it takes a long time to work. Gypsum is used in some areas of the country to make soil more acidic if the balance has shifted to alkalinity. Soil with a high pH is said to be "sweet," as opposed to acidic soil, which is considered sour. Acidic soil has pH levels that fall below a neutral reading of 7; for example, a pH of 2 is more acidic than an element that has a pH reading of 5. Basic or alkaline soil has pH levels above 7.

Fixing chemistry organically

Before you start applying soil amendments, it is time to consider again if you are seeking organic certification or at least want to use organic practices. Synthetic fertilizers boost soil chemistry but can be harmful to organisms in the soil. It is also possible you are using safe amendments, but they do not qualify for organic certification because they came from non-organic sources. For example, fertilizers made from animal waste can be used in an organic program only if it came from animals raised by organic standards. Some products say they are safe for organic production, but many do not list all their ingredients and contain substances that disqualify them from organic use. If you have questions about a product, it is best to contact your organic certifier.

Many pasture-based farmers prefer to use organic fertilizers. In many cases, the same methods can be used to spread organic fertilizers as synthetic fertilizers. Organic fertilizers can be costly, and sometimes they are hard to find. Examples of organic fertilizer include manure, raw and composted; poultry litter, manure and bedding collected from chicken houses; kelp meal, made of seaweed; and bone meal, made from ground-up animal bones (bone meal purchased in the United States is considered safe from mad cow disease). Some of these organic fertilizers come from waste generated by the animals on your farm. The availability of some organic fertilizers varies; poultry litter, for example, is most readily available if you live near a poultry house. Another option is biological fertilizers, which work by stimulating soil organisms. This triggers the digestion of organic matter, which becomes **humus**, the decayed organic matter that houses microorganisms. Humus also holds nutrients and allows the soil to hold water.

Your soil test recommendations will tell you how much fertilizer per acre to apply. Also, pay attention to the instructions on the bag because not all fertilizers absorb into soil the same way. Some fertilizers can be applied at the time of planting; others need rain to work. Many fertilizer companies will apply fertilizer for you. If you want to do the work yourself, there are a few ways to accomplish this.

Tools for fertilizing and seeding

Often, fertilizing and seeding can be done at the same time. Here are common tools used for fertilizing and seeding:

- Seeder: These tools have a hopper, a funnel-shaped bin for seeds or fertilizer. Seeds are thrown or sprayed across the pasture. Seeders come in many sizes, from those you can carry or push to those you pull behind a tractor.
- Drop spreaders: Also called buggies. These are pulled behind tractors or all-terrain vehicles (ATVs). They have a hopper to hold the seeds or fertilizer and drop the amendments directly below the buggy as it moves across the pasture.

Hand seed spreader

Seed spreader you pull behind a tractor

- Seed drills: Also called no-till drills, these tools on wheels have blades that cut the soil and drop seeds into the cuts. They also allow you to fertilize and seed simultaneously.

Seed drill

Another way to spread soil amendments is by setting up piles of fertilizer or lime and raking them over the targeted areas.

Seeding

After your soil health improves, turn your attention to the plants in your pasture. The amount of work you need to do will vary. Some farmers are able to succeed with the grasses that grow naturally in their fields. Others add the seeds of various plants they hope will complement each other by growing strong at different times of year. If your field was left untended, you will have to deal with overgrown weeds and undesirable plants before trying to improve the plant mix. If your land was previously used for row crops, you need to establish a whole new pasture.

Converting crop land

Converting a field used for growing crops into good pasture is a job that will take a few growing seasons. You have to use disks to tear up the top layer of soil. **Disks**, pulled behind tractors, have adjustable blades to break up the top soil layer. For old crop land, you will need to disk heavily to prepare the seedbed. After you seed, lightly cover seeds with soil using a harrow, a tool with spikes or disks that can be dragged behind a tractor or ATV.

Overgrown fields

The challenge with fields that have overgrown is clearing out all the undesirable species, such as thistles, a prickly plant that discourages cattle from grazing nearby. Synthetic herbicides work, but many grass-fed cattle farmers try to use as few synthetic chemicals as possible because they can contaminate the environment with harmful substances, and they are forbidden in an organic program. There are organic herbicides that contain vinegar. Mowing and hand cutting are options, as is using a blowtorch to burn problem areas. Bulldozer-type tools can be used to clear areas where trees are overgrown. After you clear the trees, you will have to seed the bare spots.

Improving pasture

Even with a usable plot of land, farmers can be tempted to simply tear up and start over so they have control over the dominant species of plants, a process called **re-seeding**. Re-seeding is not the quickest way to improve a pasture because it takes a couple of years to establish a root system for your plants, and your cows will not be able to graze on it right away. It is also expensive.

For these reasons, re-seeding is probably necessary only if your land was used for row crops or if you have cleared forest or other heavy growth. If you have established plants already, work with what you have and make improvements as you go. At first, you might not even need to seed at all.

The best thing to do to improve your field is to implement rotational grazing. *You will learn specifically about how to do this in Chapter 4.* As you manage your herd's grazing, the existing plants will be stronger, and the soil will be healthier. Improving your soil's health may make the conditions right for dormant seeds in the soil to start growing. Many of these seeds come from plants no longer growing in the pasture. If you are **rejuvenating** a pasture, which means adding new plants to an existing pasture, you will probably need about 8 pounds of seed per acre. If you are re-seeding an old crop field, you need about 20 pounds per acre.

Choosing seeds and where to get them

If you do decide to seed, talk to extension agents or seed dealers about seed mixes that do well in your area. An extension agent can be helpful because they will have a good idea of what plants work best. If you strive for the organic label, you have to be careful when considering which seeds to purchase, and make sure you only purchase organic seeds. If you cannot find what you are looking for locally, some online dealers specialize in organic seeds. Again, ATTRA, the common name for the National Sustainable Agriculture Information Service, has a database of organic seed suppliers at **http://attra.ncat.org/attra-pub/organic_seed**.

You could also do your own research and pick and choose plants to try. Sustaining your herd on pasture year-round depends, in part, on the mix of grasses growing in your field. Extend your growing season by planting grasses and legumes that hit their peaks at different times of year. This means using a mix of perennials and annuals. **Perennials** are plants that grow back from the same root systems every year, while **annuals** grow from new seeds each year. You also need both cool-season and warm-season plants and grasses so you can extend the grazing season. Cool-season grasses grow fastest when the temperature is between 50 and 75 degrees, and warm-season grasses are at their peak when the temperature is between 70 and 90 degrees.

You will want to be sure your pasture has a good mix of legumes. Legumes are plants that produce two seedleaves, such as alfalfa and clover. Legumes are high in protein and make great sources of nutrition to help you finish animals. Legumes have the added benefit of drawing nitrogen from the atmosphere into the soil, which will save you in fertilizer costs. You must be careful about giving your animals pure alfalfa too quickly because they could suffer from bloating. *Bloating will be discussed in more detail in Chapter 6.* Lower the risk by introducing cattle to alfalfa just a little each day, using an electric fence to limit their access to the alfalfa-heavy sections of pasture for a short time. Also, only introduce your cattle to alfalfa after they have eaten other grasses or hay for much of the day; if they are fairly full when introduced to lush forages, they will not gorge themselves on the alfalfa. After they gradually get used to alfalfa, usually after a week or so, their stomachs will be adapted, and it is safe to let them eat as much as they want.

Grasses and Legumes

Not all plants grow in all areas of the country or even in all parts of a state. The best way to find out which plants are suitable for your pasture is to talk to local experts such as farmers or extension agents. The USDA also has a plants database with information on hundreds of plant species at **http://plants.usda.gov**. Here is a list of grasses and legumes found in various pastures across the United States:

Grasses

- **Kentucky bluegrass**: Kentucky bluegrass is a cool-season perennial found throughout the country, but it is most common in the northeast and the north-central parts of the United States. This is a good grazing grass that can withstand repeated grazing. This grass is high quality but offers smaller yields than some grazing grasses. It can be hard to establish.

- **Smooth bromegrass**: Smooth bromegrass is a cool-season perennial. Its growing period begins earlier and continues later than many plants.

- **Sudan grass**: Sudan grass is a warm-season annual most commonly found in the western United States. It is high yielding and should not be grazed after frost because it will produce high levels of toxic prussic acid. Keep animals off for a minimum of seven days after a frost.

- **Switchgrass**: Switchgrass is a warm-season perennial. It is adaptable in much of the United States, but it is most prevalent in the Midwest. It is tolerant to droughts and floods.

- **Timothy**: Timothy is a cool-season perennial. It is often used for hay and is very nutritious when cut at the right time.

- **Orchardgrass:** Orchardgrass is a cool-season perennial. It offers high yield and high quality but does not tolerate close grazing.

- **Blue grama grass**: Blue grama grass is a warm-season perennial common in the Great Plains and Southwest. It is nutritious and can withstand close grazing. It does not grow well in areas with high water tables.

- **Bermuda grass**: Bermuda grass is a warm-season perennial that grows well in the warmest regions of the country. It is drought tolerant but not winter tolerant. It tolerates a wide range of soil pH.

- **Eastern gamagrass**: Eastern gamagrass is a warm-season perennial. It is a tall, high-yielding, nutritious plant.

- **Tall fescue**: Tall fescue is a cool-season perennial. It grows well in most soil conditions, including those that have low fertility or are acidic, and is tolerant of drought. One area it does not grow well is in the sandy soils of the Southeast. In many parts of the country, it can be infected with endophyte fungus that helps the plant survive but can be harmful or toxic to animals. The endophtye can be managed by only grazing the fescue in its vegetative stage and cutting for hay before it gets to later growth stages. The endophyte can also be diluted with forage legumes.

- **Ryegrass**: Ryegrass can be either a cool-season annual or perennial. Annual ryegrass is common in the Southeast, but it can be mixed with other pasture plants in other parts of the country. It does not tolerate cold winters. In the Southeast, the annual variety is a high-yielding plant active in winter when other plants are dormant. Perennial ryegrass is less commonly used but is often found in the Northeast and Pacific Northwest. The perennial type only lives a couple of years before it needs to be seeded again.

- **Western wheatgrass**: A cool-season perennial common in the West and Midwest, Western wheatgrass does not grow in the East. It grows earlier in the season than blue grama grass but matures later.

Legumes

- **Alfalfa**: Alfalfa is a cool-season perennial. It is considered one of the best plants for beef cattle finishing because it provides daily weight gains comparable to grains. Like many legumes, alfalfa can cause bloating if cattle are fed too much too quickly. Alfalfa grows in most parts of the country, especially in the Upper Midwest and California; it is not common in the Southeast. It grows well in summer and needs well-drained soils. For nutrient quality, alfalfa hay is the standard by which all hays are measured.

- **Birdsfoot trefoil**: Birdsfoot trefoil is a cool-season perennial. Although it is a legume, it does not cause bloating. It grows in soils with low fertility. It does better on poorly drained soils than alfalfa and grows better in July than most cool-season species. It is common throughout the United States.

- **Red clover**: Red clover is a cool-season annual. It is most common in the Northeast. It is easy to establish and is high quality.

- **White clover**: White clover is a cool-season perennial. It is easy to establish in many climates but prefers cooler, moist climates. The type of white clover known as ladino grows taller than other types. It is highly nutritious and palatable and is considered one of the best pasture legumes.

You can also use other types of plants to extend the grazing season. One option is to plant brassicas, a family of annual forage vegetables including turnips, rapeseed, and kale. Brassica plants are useful because they grow in fall and winter when other grasses may not be growing. Like legumes, these plants should also be introduced slowly because they can cause bloating in your cattle. Cattle that graze on brassicas should also be fed hay or allowed to graze pasture because brassica plants are low in fiber, essential for rumen health and productivity. Herbs can also be used for grazing. One herb great for grazing is chicory, which can produce weight gains similar to alfalfa. It grows well in spring and summer.

Universities in your area or the state Natural Resource Conservation Service offices publish plant guides that tell you what plants work in your area, what plants work best in mixtures together, when to plant, and how many pounds of seeds per acre to plant. These guides also include the best times to harvest these plants for hay. The keys to success in the seeding process are the timing of planting, the fieldwork you do to ensure seed contact with the ground, and choosing the right seeds.

When to plant

Generally, the best times to plant are spring and fall. If you plant in the fall, do it early enough that the seed has time to establish itself before the first frost. You do not want the seed to germinate right before the first freeze because it will not survive. If you plant in the spring, plant early enough that the plants can establish roots before the heat of summer.

The exact timing of these dates depends on where you live. It takes plants between one and two weeks to germinate or begin to grow, and the plants probably need another month to establish root systems strong enough to withstand the winter. The National Climactic Data Center has a searchable database of temperature and precipitation data, including charts that break down the probable dates of first and last freezes in each state, and in different areas of each state (**http://cdo.ncdc.noaa.gov/cgi-bin/climatenormals/climatenormals.pl**). Experienced farmers or local extension agents also can give suggestions on the best times for fall planting in your area.

If you are already using your pasture for grazing, you will not be able to seed the whole field at once. You will have to keep your herd off the seeded sections so they do not harm the new plants. Cows do not bite grass, they pull it, and if the roots are not well developed, cows will pull the whole plant out of the ground. The exact height of grass varies by species, but you want your new plants to be about 6 or 8 inches tall before you allow your cattle to graze on them.

Before you seed, either mow or let your cattle graze down the competing plants so they do not smother the new plants. In this case, overgrazing is permissible — if you overgraze the existing plants before the onset of winter, their spring growth will start slowly and give the new seeds a better chance to grow successfully. Planting in the fall gives new grasses a better chance to establish themselves before the spring, when the older plants could grow so fast they would smother the new plants. If you plant in the spring, you will have to monitor the established plants to be sure they do not smother the new ones. If these established plants grow too quickly, cut them for hay.

Seeding strategies

Seeds do not have to be planted very deep. Depending on the species, a depth between ½ and ¼ of an inch is usually sufficient. Plant your seeds when moisture is available. You can plant seeds using either a broadcast seeder or a seed drill. A **broadcast seeder** is a machine used to apply fertilizer or seeds; it spreads seeds by casting seeds outward in many directions. A **seed drill**, also called a no-till drill, is a more precise planting tool. These drills have blades that cut the soil and drop seeds into the cuts to ensure good contact with the soil. If you use a broadcast seeder in spring or fall, lightly disk the pasture first. Disking is not necessary with a seed drill because they have blades that cut the soil and drop the seeds into the cuts.

Farmers often use a technique known as **frost seeding**. To frost seed, farmers broadcast seeds onto pasture in late winter or early spring while the ground is still frozen. The idea behind frost seeding is the freezing and thawing of the ground creates cracks in the soil that allow seeds to penetrate the soil without using a seeder drill. Another way to distribute seeds is to let your livestock do it for you by mixing the seeds in the herd's mineral supplements. Animals eat the seeds and distribute them in their manure. The animal's stomach acids may help break down the outer seed shell to increase the chances the seeds will grow properly. This is an imprecise way to spread seeds, but it is comparable to the way animals spread seeds in nature.

Irrigation

Another important item to consider when planting your pasture is irrigation. Many farmers turn to irrigation when there is no rainfall for their pastures. **Irrigation**, artificially applying water, can be an effective tool for pasture management. There are places, in the West especially, where irrigation is vital. However, irrigation's costs reduce profit margins, so irrigate pastures only if necessary and as efficiently as possible.

Well-managed pastures reduce the need for irrigation. Healthy plants impede the flow of water, which gives healthy soil more time to absorb it. Healthy plants also shade the soil and prevent it from drying out, which is helpful because moist soil absorbs water better than dry soil. A healthy pasture experiences less evaporation and runoff.

In areas where irrigation is necessary, such as in Southwestern deserts, areas in the valleys around the Rocky Mountains, or some prairies in the Great Plains, water can come from both on-farm and off-farm sources. On-farm sources include streams, lakes, ponds, wells, drainage ditches, reservoirs, or even recycled irrigation runoff, while off-farm sources are usually suppliers, including cooperatives, community water systems, or commercial companies. About 20 percent of farmers in the West get their irrigation water from the U.S. Bureau of Reclamation, a government water-management agency that produces hydroelectric power and is a wholesaler of water. Off-farm water could include rivers and lakes, or reclaimed water, wastewater treated for non-potable or non-drinkable reuse.

Irrigation systems are divided into two types: gravity-flow and pressurized. If you live in an area where pasture irrigation is common, local farming experts can give you an idea of what might work best for you. Here is a look at irrigation types and different systems, adapted from information about irrigation on the USDA website at **www.ers.usda.gov/briefing/wateruse/glossary.htm**:

- Gravity-flow systems channel water onto fields through ditches or pipes or use gates to release water onto fields.
- Pressurized systems include sprinkler and low-flow irrigation systems. Sprinklers work better on rolling terrain than gravity-flow systems because water will find ways around hills and reaches high spots that gravity-flow systems can not reach. Pressurized systems also usually use pipelines for their water supplies.

Gravity-flow systems

The following are examples of gravity-flow systems you can use on your pasture:

- Flood systems, also called border systems, release water onto fields through ditches or pipes. The pipes funnel water that comes from natural sources such as rivers and lakes or from man-made reservoirs, and the water simply washes over the field.
- Furrows are another type of irrigation system, but they are usually used for row crops. Instead of letting water wash freely over a pasture, furrow channels steer water between crop rows.

Sprinkler systems

There are several different options for types of sprinkler systems you can use on your farm:

- Solid-set sprinklers are stationary and often permanent. These systems use aboveground sprinkler nozzles that spray water pumped through supply lines fixed either above or below the surface.
- Hand-move sprinklers are portable sprinkler systems you must move by hand from one area of pasture to another. It requires a lot of time to move these systems, but they do roll for easier movement, and they can work for small fields.
- Side-roll wheel-move systems are similar to the hand-move models, but they are powered by gasoline engines to make them easier to move.
- Big-gun sprinklers are self-propelled sprinklers mounted on a wheeled cart or trailer. The force of the water spray pushes the sprinkler through the pasture. These high-pressure systems can also be used to spread livestock waste.

CASE STUDY: ROCKY MOUNTAIN ORGANIC MEATS

Rod Morrison
Chief operating officer and president
Rocky Mountain Organic Meats
Powell, Wyoming
http://www.rockymtncuts.com
info@rockymtncuts.com

For Rod Morrison, raising grass-fed organic beef is not just a business; it is a crusade. Morrison has always believed in farming the old-fashioned way, as organically as possible. He is part of a farmer-owned meat company that raises cattle without hormones, antibiotics, or synthetic fertilizers. The land ranched for the company includes 600,000 acres of the Arapaho reservation near Yellowstone National Park in Wyoming. Morrison is a follower of Wendell Berry, the writer who champions sustainable agriculture, stewardship of the land, and the value of rural communities.

Morrison hopes more producers and customers realize the benefits of organically raised grass-fed cattle. He argues conventional agriculture uses energy at an unsustainable rate. Morrison's goal is to use no more energy each day on his farm than what reaches the Earth as sunlight.

"Honestly, it is the only way we are going to be able to sustain ourselves in the long run," Morrison said. "You cannot go halfway. That is like being half married. If you are half married, then good luck."

He is the president and chief operating officer of Rocky Mountain Organic Meats, a producer-owned company of between ten and 15 farmers who raise grass-fed organic beef and lamb near the Rocky Mountains in Wyoming. All the producers are certified organic according to National Organic Program standards. This certification is important to the company's producers because they want their customers to be confident of what they are buying and of where it came from.

He ships boxes of meat to customers all over the country. But he does not just take orders. He wants to communicate with his customers and ensure they understand why he raises his beef organically and why it is important they choose such food. Most of his boxes go out weighing 20 to 30 pounds, but he also offers a package that equals a side of beef — roughly 175 pounds of steaks, roasts, burger, and specialty slices — for $1,400. He charges about $7 a pound, a price most customers will not pay but is worth it to customers who understand the benefits of raising cattle organically on pasture.

"A few are [discovering the benefits] every day," Morrison said. "And as they figure it out, they are understanding now they have a relationship with someone who produces food, and they are very happy about that."

Morrison's farm is nestled in the Big Horn Basin at the base of Heart Mountain. His fields used to be raised for sugar beets and malt barley. He planted everything with a mixture of alfalfa and cold- and warm-season grass seeds. He let these plants establish themselves for a year and a half before he began using the fields for grazing. Morrison raises 45 mother cows that produce calves each year. He also buys about 40 or so calves each year from another farmer who raises his cattle organically but does not want to finish them. Those calves arrive at about 600 pounds and are grown to between 900 and 1,100 pounds. His cattle usually finish in about 18 to 20 months.

• Chapter 3 •
Facilities

TERMS TO KNOW:

Post-hole digger: A clamshell-shaped tool used to dig holes for posts to be placed in

Augur : A drill-like tool attached to a tractor or a skid steer

Stanchions: Devices attached to a cow's head to keep it still during milking

Tie-stalls: Tools used to hold cows during milking

Bucket system: A system of milking where vacuum tubes pump milk into buckets and the milk is poured from the buckets to a holding container

Bulk tank: A stainless steel refrigeration unit that keeps milk cool until it is collected

Plate cooler: A device that pre-cools milk before it gets to the storage tank

The best advice for someone who is about to invest in the infrastructure for a grass-fed cattle operation is to keep it simple. You might be tempted by all the neat machinery and products advertised as making life easier, and most of them work according to the description, but their price tag could put your farm in jeopardy. Many farmers fall into the trap of having to continually expand to survive. You must consider each purchase carefully to be sure it is necessary. You might be surprised how few facilities you can get by with.

If there is not a barn on your property already, you will probably be fine without one. One of the key principles of grass-fed cattle farming is cattle should be left outdoors, as they would live in nature. Many dairies are being built without winter housing. The best approach probably is to start small and give the minimalist approach a chance to work. You can always add more equipment and facilities down the line if you decide you cannot do without them.

Cattle have basic needs for which you must provide. They need to be shielded from the wind and the sun, which can be accomplished with trees or hills. They also need health care. *This topic will be discussed further in Chapter 6.* They probably need fences to keep them from getting hurt on the highway and to prevent them from eating your neighbors' yards. Some Western states have open-range laws that permit cattle to be unfenced, but the majority of grass-fed cattle farmers use electric fencing. Laws vary by location, so start with county zoning officials, or your local extension agent, to see what the rules are for you. You will have to be sure your cattle have access to water in your managed grazing plan. *You will learn more about water-source options in Chapter 4.*

It is important to design your facilities so your animals are comfortable. Here are some tips to follow so your cattle are happy in their new facilities.

- When you bring them into a holding area, give them enough space to follow one another and maintain visual contact with the leader cow.
- They do not like to go around corners unless they can see where they are going. Make sure turns are gentle and rounded rather than sharp and angular.

- Bright lights, shiny reflections, and shadows alarm cattle, as do loud noises such as clanging gates. Inspect your facilities to eliminate these distractions.

- Be sure your fences and pens are well maintained. Animals can hurt themselves on nails, sharp corners, or broken boards.

- Provide sufficient lighting because they will not want to enter a dimly lit milking facility and will hesitate to walk up steps until they are allowed time to investigate.

If you do have to house an animal for an extended period of time, it is important to have a well-ventilated facility and to provide clean bedding. Bedding materials include sawdust, ground corncobs, straw, or wood shavings. The depth of bedding should be 10 to 12 inches or more. Be sure the bedding stays clean and dry by removing soiled areas and re-covering those areas with new bedding.

Holding Pen

If you are starting a farm from scratch, plan your holding-pen area first. You need a holding pen for routine herd management tasks, for treating sick animals, or to receive new animals. A holding area is a pen where groups of animals stand while they wait to be moved into another area, such as a loading chute or health care area. The ideal holding pen is round with enough space for each confined animal to turn around. This holding area is usually attached to an alleyway or chute, basically a panel on either side of the cow, leading to a head gate that restrains the animal's head during health care procedures. The holding area could also lead to a loading chute for cattle leaving the farm. You need an entrance gate that leads from the pasture, a gate leading into the chute to the head gate, and a gate to release the animal back into the pasture after treatment.

The walls of your holding pen should be at least 60 inches tall, and it should be made of boards, plywood, or some type of solid, non-transparent material, such as steel. When the cattle are in the pen, in alleys, or in chutes, they should be able to see where they are going, but they should have minimal visibility to what is going on outside their area. Cattle traffic should flow easily both ways between your holding pen and your pasture. The pen should be accessible in all weather by truck. You also need a power source for water or for light to take care of sick animals at night.

Your chute leading to the head gate can be short or long — a short one works if you bring animals in one at a time, and a long one allows more than one animal to wait in the alley. The holding pen and the chute should be round because cattle naturally like to circle back and corners will cause them to bunch up. Mississippi State University has a page of facility design ideas at the following address: **http://msucares.com/pubs/plans/books/beef.html**. Dr. Temple Grandin, a professor at Colorado State University, also has a great site for pen designs at **www.grandin.com**.

If you do not want to build these facilities, you can buy pens, chutes, and head gates at most farm supply stores. A couple of well-known chains are Southern States (**www.southernstates.com**) and Tractor Supply Co (**www.tractorsupply.com**). The alleys can be simple, narrowly placed fencing, or they can be squeeze chutes that close in to hold an animal's sides so it cannot move. A manual head gate will cost a few hundred dollars; a squeeze chute system with a gate that moves with a hydraulic mechanism can cost a few thousand dollars. A hydraulically working chute with all the options could cost as much as $10,000.

You may use fencing to create a pen to receive new animals if you routinely buy them. This is a good option if your farm focuses on stocker cattle. This receiving paddock may need a sturdier fencing type than a simple electric fence, such as mesh or barbed wire, because these new animals may not be accustomed to electric fencing. They will also be stressed from the move and may naturally have a wild temperament. This quarantine area allows you to watch them for signs of illness before bringing them in with the rest of your animals. You will also be able to process new arrivals with vaccinations. Another use for fencing is to create a separate area to isolate sick animals. These areas need a water source and shade.

Storage

If your farm already has a barn or other storage facility, this is a safe place to store hay. Alternatively, you can build three-sided facilities. If you use a facility, make sure to keep the bottoms of your bales dry. One way to do this is to set bales on wooden pallets to keep them off the floor. Again, you do not necessarily need a barn to store these materials. If you do not have a barn or shed, you can store hay outside if you cover it with plastic or a tarp. You may also use silage pits to store a certain type of food for periods of low forage quality, such as in winter. **Silage** is fermented, moist forage made from almost any green, growing plant. It is stored either in silos or in concrete bunkers covered with air-tight tarps. If it is not stored properly, a dangerous bacterium called *Listeria monocytogenes* could sicken or kill your animals. *You will learn more about harvesting and protecting feed in Chapter 7.*

Shelter

Cattle have lived outdoors for thousands of years, and they can survive even the most brutal climate extremes. However, they do need some form of relief from both cold winds and the sun's rays. Overheated livestock feel malaise, and they will not eat. High winds stress animals, and they need more energy to keep warm in the cold. Barns or other existing structures could provide shade or relief from wind. Your geography may provide some natural relief for your cattle. Tall hills or trees already on your property can slow the wind and block the sun. Some farmers also plant rows of trees as windbreaks, but it could be a few years until they grow tall enough to provide any shelter or protection.

Windbreaks

Many farmers construct artificial shades or windbreaks out of aluminum, steel, or other sturdy materials. These structures can be one-sided, meaning they have one wall and an overhanging roof held up with two posts, but some farms use simple open-sided sheds to protect animals from the harshest weather conditions. Portable units with wheels are especially useful in a rotational grazing system. Move these units around to encourage the animals to distribute their manure and urine around the pasture. Where you place your shades or wind breaks depends on your location and the direction from which cold weather rolls in.

Fences

Fencing can be one of the most costly expenditures for a new farm. For a rotational grazing program, you need permanent and temporary fencing. The type of permanent fencing you use is a matter of personal choice. A common choice in rotational grazing systems is electrified, high-tensile wire as a permanent fence and portable electric wire for temporary paddock divisions. *Rotational grazing systems will be discussed in more detail in Chapter 4.* Use different fencing types for different sections of your farm; for example, high-tensile wire on one section of the farm, and barbed wire in another section. The most common types of fencing material are:

- **Barbed wire**: Barbed-wire fencing has sharp barbs to discourage animals from trying to cross through the fence. These fences, which usually consist of three to six strands, are strong and require little maintenance because once they are up they will last a while and are not as prone to sagging as electric wire fencing. However, barbed wire is not merciful to wildlife that gets tangled in the fence. It should never be electrified because animals — or people — could get hung on the barbs, which could be lethal.

- **Electric wire**: Use two or three strands of electrified wire stretched across posts 50 or more feet apart. This is the cheapest and easiest fence to put up and can be used as permanent fencing. Electric wire is not a physical barrier but a psychological one. Cattle can easily go through electric fencing, but if they are shocked, they will not want to. Electric wire fences can be hard for cattle to see, but if you turn them loose in a field they will stumble upon it. After this, they will be able to sense the current when they are near one.

- **High-tensile wire**: High-tensile wire fences are easy to install and stout, meaning they can withstand impact from large animals without snapping. Farmers usually use two to four lines of high-tensile wire, and usually two of these wires are electrified. This type of fencing requires regular maintenance to ensure wires do not sag and weeds or tall grasses do not make contact with a line and drain its current.

- **Wood rail**: Wood rail fences are a highly visible barrier for cattle. These fences are expensive and require a lot of maintenance, as you have to keep the wood treated and painted.

- **Woven wire**: This type of wire is mesh fencing that is high strength and low maintenance. It is the most expensive fencing type. Many farmers also add a top strand of barbed or electric wire to prevent cattle from rubbing against the fence.

  Purchase wire fences in rolls of a couple hundred to a couple thousand feet. Check with your local building officials to ensure you install a legal fence. There may be height or post-spacing requirements, restrictions near roadways, or other rules you must follow. Be sure to install fences on your side of the property line and not your neighbors'.

  Generally, permanent cattle fences are 48 to 54 inches tall. Fencing systems also need end posts, line posts, brace posts, and gates.

- **End posts** are the cornerstones of your fence line. They are usually made of sturdy wood or steel. They must be buried deeply, about one-third to half of their total height below the ground, to withstand the pressure of being pulled by taut fencing. You can also pour cement in the hole around them for more support.

- **Brace posts** provide support for end posts. They are usually placed 6 to 8 feet away from the post being braced. An H-brace is two parallel posts with a horizontal post placed in between. Use a brace wire from the top of the brace post to the bottom of the post being braced. You could also use a diagonal brace wedged between the ground and the end post, which requires a brace wire. Another option for lower-tension fences is a bed log brace, simply a log buried on the tension side of the post that prevents the post from being pulled over.

The University of Wisconsin–Riverside has a good guide to bracing posts at **www.uwrf.edu/grazing/bracing.pdf**.

- **Line posts** are needed to support fencing during long, straight stretches. You can buy line posts made out of wood, steel, fiberglass, or plastic.
- **Gates** need to be as wide as the animals or the machinery you plan to move through them. You can get gates as small as 4 feet, good for people, or as wide as 16 feet, for large machines. One that is 12 to 14 feet wide will work for most tractors. Gates come in light, medium, and heavy weights. The heaviest gates are called bull gates. You would need a heavy bull gate in the sorting and handling areas near your chutes. Lighter gates should be fine if used in areas that have lightweight electric fencing.

It is also important to note that under the National Organic Program, treated wood is not allowed for fence posts or for any structures that come in contact with animals or crops. Alternatives for fence posts include naturally sturdy wood such as black locust that resists rot, or you can use alternative finishing products to treat your wood. Your organic certifier will be able to tell you what you can and cannot use. ATTRA provides information about sustainable agriculture and has a good discussion on the subject at **http://attra.ncat.org/attra-pub/lumber.html#usda**.

Tools you may need to install a fence

The following represent some of the tools you need to build your own fence. Note that your needs may change depending on the type of fence you plan on building.

- Posts can be put in the ground manually with a **post-hole digger**, a clamshell-shaped tool used to dig, or with an **auger**, a drill-type tool you attach to a tractor or a small farming vehicle called a skid steer.
- Stapler and staples 1 to 1½ inches long to attach fencing to wooden posts
- Wire stretcher or fence puller to tighten fences
- Hammer to tap smaller posts or rods into the ground
- Spinner or reel that allows you to roll or unroll wire easily
- Round crimper, a tool for making good electrical wire connections
- Digital volt meter that tests grounding systems and helps you ensure each section of line has sufficient charge throughout the system

Installing your fence

Design the layout of your permanent fencing system by using an aerial map of your farm as a guide. These aerial photographic maps are available for free at the local office of the Farm Service Agency, a division of the USDA that helps administer farm programs. This aerial photo map allows you to mark your property lines and to plan around existing buildings, water sources, and shaded areas.

Install a fence yourself, or hire a contractor to do it. Here are some tips for putting up fence yourself:

- If you put fence posts up by hand, wear thick gloves and eye goggles.
- Call your local utility companies to check for buried lines before you dig.
- Dig deep enough that the post will not lean; this is usually about one-third the height of the fence. Virginia Tech has a good guide to installing high-tensile wire fences that includes good instructions

for end posts and brace posts at **http://pubs.ext.vt.edu/442/442-132/442-132.html**.

When installing a fence, put up your end posts first. Run a temporary wire from post to post to help keep your line posts straight. The spacing between your line posts will depend on the slope of the land and the type of fencing you use. (For example, space posts further apart on flat land, and the heavier the fence type, the closer together the posts need to be.) Posts between woven wire need to be closer than posts for other fencing types because woven wire is heavier and needs more support; posts for woven wire should be about 10 to 14 feet apart. Line posts for barbed wire should be 14 to 16 feet apart. Posts for high-tensile wire, which is lighter than barbed wire, can be 15 to 20 feet apart, and posts for lighter electric fences can be spaced at least 40 feet apart on flat land.

Gates should always go in corners because it is easier to funnel cattle toward corners than toward a gate in the middle of the fence line. Avoid low spots because cattle like to walk uphill. Gates should be wide enough to provide access for farm equipment or trucks. Add a smaller gate to allow access so people do not climb the fence or the large gate.

Components of an electric fence system

Here are terms you may encounter as you develop an electrical fencing system:

- **Wires**: For permanent fences, use a thick wire because they carry current farther than thin wire. The USDA recommends 12.5 gauge or higher. Temporary fences can use lighter, six-strand polywire or polytape.

- **Posts**: Your fence will need end posts, which can be wood or fiberglass. Line posts can be T-posts, which must be pounded into the ground. Step-in posts, which feature a flange you step on to push them down, are much easier to put in the ground and are a good option for temporary fencing.

- **Insulators**: These let you attach electric wires to posts without actually touching the posts so the current does not travel through the post into the ground. Good plastic insulators work just fine.
- **Crimping Sleeves**: These help you join two wires together. You put both lines inside the sleeve and use a crimping tool to squeeze the two lines together.
- **Line tap**: Line taps connect live wires to existing wires. They come in two styles, crimping and split bolt. Crimping-type line taps make permanent connections by splicing lines. Split bolt taps work for gates or other sections where you may not want a permanent connection.
- **Line switches**: These switches allow you to turn off current flowing through sections of fence instead of the entire line.
- **Gates**: Gates have handles that latch onto an anchor connected to a post. When they are unhooked to allow passage, these cause the wire to go dead. Instead of gates, some farmers with smaller herds carry a PVC pipe with a notch on the end, and they use the pipe to lift or lower the fence to allow passage.
- **Insulated wire**: Bury insulated wires under your gates so you do not have to turn off your fence every time you use the gate. The USDA recommends 12.5 gauge wire. For added protection, put the insulated wire in a PVC pipe. Dig a trench at least 12 inches deep, and then cover up the line.
- **Chargers**: Also called energizers, chargers send the current through the system. The USDA recommends a low-impedance energizer with a minimum 5,000-volt output. Low impedance means low resistance

to electrical current. Another important measurement is joules, which measure the strength of the shock. The higher the joule, the greater the shock. Chargers can be powered by alternating current (AC), battery or solar energy, or dual battery/AC.

- **Ground rods**: These rods collect the energizer's electric circuit and provide the shock when something touches the fence. Ground rods should be installed near the charger, and they should be at least 6 feet long. Use as many ground rods as recommended by the energizer manufacturer.
- **Lightning arrestors and surge protectors**: These protect the charger from electricity damage. Surge protectors keep the charger safe from common current surges, and lightning arrestors divert lightning strikes away from the charger.

Electric fencing is either used as the sole component in fencing systems or combined with other fencing types for added protection. It is also ideal as a movable interior barrier in rotational grazing systems. *Rotational grazing systems is discussed in Chapter 4.* Electric fencing is fine to hold cattle so long as they are not crowded or spooked. Cattle are curious and will probably discover the fence on their own; when they do, they will leave it alone because the bite of an electric fence hurts. Another strategy is to introduce cattle to the electric fence. The best way to do this is to stretch a live wire across a small paddock, such as your receiving paddock, and entice the cattle to contact it. Placing the line between the cattle and a water source, or the cattle and hay, is another effective strategy for introducing your cattle to electric fencing.

Installing electric fencing is something you probably can figure out yourself. To stay safe while putting up your fence, wait until the whole system is up before you plug in your fence charger. Check your area's laws about posting warning signs on electric boundary fences and warn neighbors, children, or visitors about the fence. If you have never installed an electric fencing system before, you may want to seek advice from a trusted expert such as an experienced farmer or a local extension agent.

Electric fences work as an incomplete circuit from the charger through the fence system. The circuit is complete when it is contacted by an animal, person, or anything else touching the ground. You have to check your fence frequently to make sure the current is flowing through each section and that no wires have come down. A visual inspection will detect downed lines or lines in contact with tall weeds. This saps the current, so you will have to be sure the areas under the fence are grazed or mowed down. A digital volt meter will further help you check your line for problems.

Dairy Farm Facilities

Dairy farms can be as small as one cow or as big as herds with more than 1,000 animals. Several different kinds of equipment and methods are used to milk these animals, from hand milking and simple bucket systems to complex mechanized milkers. The bucket systems can be labor-intensive but work fine for a small number of animals. The rotary systems, carousels that rotate cows around the farmer who stands in the center of the parlor, are expensive but allow you to milk massive herds quickly.

If you do not have much equity, it may be better to get experience by working on someone else's land while you build your herd, to lease someone else's farm, or to purchase an old dairy farm so you can get your milking equipment when you purchase your land. The dairy market has been depressed, so it is possible to buy good used equipment on sale when a dairy goes out of production. Joe Horner, an extension professor for the University of Missouri College of Agriculture, Food, and Natural Resources, suggested an idea he saw Mennonites implement to purchase used dairy systems: They go to public sales of farms that are closing down and buy everything for much cheaper than what it would have cost used. "Mennonites are famous for this," Horner said. "They will come in and take [the dairy system] down piece by piece, get permanent markers out, write which piece goes where, and reconstruct it in their own barn."

Milking parlor types

If you are milking a small number of cows, you may be able to start out with a couple of homemade stalls and a portable milking machine. If you are buying an existing dairy barn, you could use the existing equipment or install new, more efficient equipment in old areas. Old dairy barns often use stanchions or tie-stalls to hold animals during milking. **Stanchions** are devices that latch around a cow's neck to secure it during milking, and **tie-stalls** use neck chains to hold cows during milking. These facilities usually use bucket milkers or pipeline milking systems. In a **bucket system**, vacuum tubes pump milk into a small bucket, and the milk is poured from the bucket into a holding container. In these systems, the operator is on level ground with the cows, so you have to bend to attach the teat cups of the milking machine. If you use a bucket system, you must also bend and lift to carry the bucket to the holding tank, or you can pour the milk through a filter directly into bottles. Pipeline systems pump milk from the milker to a holding area, which eliminates the buckets and the need to carry them, but the farmer must still bend to attach the teat cups to the cow. These systems are labor-intensive but can be effective for small-scale farmers. Some stores specialize in equipment for small dairies, such as Family Milk Cow Dairy Supply Store (**www.familymilkcow.com**), which offers such equipment as milking machines and cheese-making tools.

Newer dairy systems rely on milking parlors to reduce bending and lifting by elevating the cows above the milker, who stands in an area called the operator pit. This cuts milking time and leaves time for other management tasks. The smallest dairies may not find these systems to be cost effective. Just 22 percent of small organic dairies that had 50 cows or fewer use them, according to the USDA. But nearly all organic dairies with 200 cows or more use them. Many farmers will update existing barns with newer parlors. Some parlors are more basic than others. In most milking systems, the herd, or a portion of the herd, is walked to the parlor and held in a holding pen before each milking. Cattle wait behind a crowd gate for a stall to open.

Here is a list of milk parlor types that could work in a grazing-based system:

- **Herringbone parlor**: Cows stand side-by-side, angled toward the operator pit. Milking clusters can be attached to the teats from the side of the cow, which makes it easier to see and clean front udders.
- **Parallel parlors**: Similar to herringbone parlors; cows are milked from the rear between their hind legs, which provides less udder visibility than the herringbone design. The walking distance between cows is shorter than in the herringbone design.
- **Side opening parlors**: Cows stand head to tail in these stalls. The walking distance in this system limits the number of stalls you can use in this design, but some farmers prefer this type of milking system because the entire cow is visible and can be checked for health.
- **Rotary parlor**: In this system, the milker stands in one spot while cows are rotated around as if on a carousel. This system is efficient, but labor intensive because cows are rotated by in a matter of seconds. It is probably only cost effective on very large dairies.
- **Swing parlor**: In this system, two groups of cows stand in the parlor, one on each side of the operator, and one side is milked before the operator swings the hoses to the animals on the other side. When not in use, the milking hoses hang in the operator area. Cows are milked

from the rear between their hind legs. Swing parlors are a popular option now because they are cheaper to install than other parlor types but still increase the number of cows milked in an hour.

- **Walkthrough parlors**: Cows enter from the rear, are milked on a platform, and exit forward through a gate.

Most parlor types are safe for both cattle and the farmers if the cows are properly handled and the farmer is properly trained. When choosing a parlor type, farmers weigh cost against efficiency and even personal preference on the way cows stand in the parlor. Rotary parlors milk the most cows in the shortest amount of time, but the cost is usually prohibitive for herds of fewer than 700 cows. Automated, rapid-exit parallel parlors are probably the next fastest, and are common in herds of 300 or more. For many new dairies with 300 cows or fewer, swing parlors seem to offer the best compromise between cost and efficiency — they cost about half as much as herringbones but still provide about 75 percent of the speed.

You want the entire process for each milking to take less than two and a half hours because you want your cattle to be out in the pasture where they can eat and rest. In really hot weather, many farmers delay the day's second milking until the evening when it is cooler. You can also use fans to cool cattle. Some dairy farmers invest in mist cooling systems to keep the cows comfortable because heat-stressed cattle produce less milk per day than comfortable cows. Milk production can drop about 2 pounds per cow on days when temperatures top 90 degrees, according to University of Nebraska-Lincoln researchers.

Milk-cooling equipment

If you plan to have a small operation, your cooling system may not be much more complicated than bottles and refrigerators. Many dairies use pipelines to transport milk from the parlor into the milk house, where the refrigerated storage tanks are kept. The temperature of milk is about 101 degrees when it leaves the udder and must be cooled to 38 to 45 degrees within two hours of milking to keep it fresh. A farmer with a small herd who sells raw milk off the farm could pour milk through a filter straight into bottles customers take home. Most small farmers probably just need to store their milk in a **bulk tank**, a stainless steel refrigeration unit that keeps milk cool until collection. These tanks come in sizes from about 600 gallons to about 8,000 gallons. A common type of cooling equipment on large dairies is a **plate cooler**, which pre-cools milk before it gets to the storage tank. The warm, fresh milk flows on opposite sides of steel plates from cool water, transferring its heat to the water. This saves electricity that would have been used to cool the milk in the holding tank.

Permanent lanes

Permanent travel lanes take on extra importance on grazing dairies because cattle are moved from the field to the barn once or twice a day for much of the year. These lanes are necessary because the cow procession is tough on the land and leads to compacted, muddy lanes. You want your travel lanes to be wide enough to move the whole herd, but there is no set formula for this. Some farmers march their animals single file like ants down 2-foot-wide lanes. Often, farmers go with 24-foot-wide lanes to also allow access for tractors or other equipment.

Cows do not like walking around sharp corners, so you need gentle, looping corners. A good travel lane will be made of firm, flattened material such as clay, crowned in the middle like a road. It should be covered with a soft, safe, natural material such as lime so the animals' hooves are not injured during the walk. Find materials with these general qualities native to your area because it is expensive to truck in materials from other areas of the country. When designing the lanes, use an aerial map but also walk the farm to be sure there are not unexpected hills or other challenges. Match the topography so your lane does not impede the natural flow of water; otherwise, you will have to add culverts to allow water passage. Try to plan the lanes with natural shade so the animals do not have to make the trek in the heat.

Handling manure

Manure collects in milking areas and must be removed. If you have a small number of animals, just shovel the manure into a wheelbarrow and take it to a compost pile. Larger dairy barns have slots in the floors behind the animals so manure will fall into a pit beneath the floor of the barn where it can be collected later. Some dairies flush the floor with water. You can also scrape the manure with an attachment for a skid steer. This manure can later be used on your field as compost or sprayed directly on pasture.

Other Equipment You May Need

The equipment you invest in depends on what chores you want to handle yourself. It is best to keep your investment in equipment low, but there are many chores on the farm you will not be able to do without machinery. You can probably get by with renting some of this equipment or using a contract laborer who has this equipment and will do the work for you. Keep in mind that contract labor can cost about $30 to $60 per hour.

You will occasionally need to haul heavy loads or pull other machinery, so you will need a dependable vehicle on your farm. You could purchase a tractor, but these are costly and their value depreciates quickly, or you can rent a small tractor from a machine rental business. If you do not own a tractor and do not want to buy a pickup truck, small-scale farmers can get by with an all-terrain vehicle that can pull small trailers, manure spreaders, and harrows. You can also get front-end attachments to plow snow or rent a skid steer, a versatile machine that comes with attachments for many farm chores. These attachments include augers for fence postholes and hay bale spears. Use one of these attachments to scrape manure from dairy floors.

Many farmers have also invested in the equipment needed to make hay. However, making hay is an expensive process that requires cutting, raking, and baling. When you are first starting your farm operation, it might be better to buy hay, share equipment with a neighbor, or arrange for a contractor to make your hay for you. *Hay will be discussed in more detail in Chapter 7.*

Another chore you must consider is hauling animals to and from your farm. If you will be transporting animals frequently throughout the year, it may be worthwhile to own a stock trailer, a trailer used for hauling cattle. You would need a truck to pull this trailer. If you will be making a limited number of trips, it may be more cost effective to hire a custom livestock hauler who will haul your cattle for you for a small fee.

When choosing to invest in facilities or equipment, it is smart to look around for cheap alternatives. The less money you tie up in equipment or buildings, the more flexibility you have in your budget, and the better your chances are of making it in the long run.

In the next chapter, we will look at ways to get the most out of the cheapest cattle food source — your pasture.

CASE STUDY: FULL CIRCLE FARMS

Dennis and Alicia Stoltzfoos
Owners
Full Circle Farms
Live Oak, Florida
thisisdennis@juno.com

Eight years ago, Dennis and Alicia Stoltzfoos started their farm with one cow and her 6-month-old calf. Back then, they were not sure anyone would be interested in their products, and Dennis Stoltzfoos was not thrilled that milking was an everyday chore. Today, Full Circle Farms is a beef, dairy, poultry, and pork operation that grosses $200,000 per year. The business succeeds for the same reason Dennis Stoltzfoos was willing to do the work — customers can not find more nutritious products anywhere else.

Dennis Stoltzfoos, the youngest of 11 children, was raised Amish/Mennonite on his father's conventional farm. He flirted with various jobs and spent 15 years in the alternative health field. In his early 20s, he was constantly bothered with various aches and pains, low energy, and allergies, and he grew frustrated after a battery of tests revealed nothing wrong. His doctor told him his problems were in his head. So he started doing his own research and became a disciple of Weston A. Price, whose 1939 book *Nutrition and Physical Degeneration* demonstrated the benefits of animal fats and other natural foods versus processed modern diets.

"Your body makes 300 million new cells every minute," Stoltzfoos said. "You determine if those cells are healthy, medium, or weak by what you put in your mouth. Our goal is to make the most nutrient-dense products possible. We were not sure in the beginning if people would pay for food produced like this. The most surprising thing is that people cannot get enough. You cannot make it fast enough."

When Stoltzfoos re-committed to the year-round job of farming, he embraced the idea of sparing no expense to have the healthiest grasses for the healthiest cattle, which equals the healthiest products. Their milk is in such demand that they charge $13 for a gallon of raw, unpasteurized milk. The prices other farmers charge varies, but $10 or less is common. They have about 100 head of cattle, including 30 Jersey milk cows. The family owns about 60 acres and leases a couple hundred more. Stoltzfoos uses creativity in his management decisions and borrows principles from Joel Salatin, the famed Virginia farmer who preaches the benefits of using every animal as a tool for the good of the whole. For example, the farm's chickens move into paddocks when cattle move out so the chickens can scratch for fly larvae in the manure, which aids in soil fertility.

Stoltzfoos built his own milking stall for $100 with some 2x4 and 4x4 wooden boards. In the winter, he entices the cows in with alfalfa pellets or molasses in a trough; in summer, he removes the trough. He never feeds his cattle grain. He uses a milking unit from an Amish company called E-Z Milking Equipment. It is a vacuum pump with two buckets and two milkers. He empties the bucket through a strainer into a 5-gallon container with a tap, and his 8-year-old daughter uses the tap to fill 1-gallon jugs for customers. They spend about one hour per day milking.

They bought their first cow and her calf before they had land and held her at a friend's farm. They leased land for a couple of years before buying their own farm. As their farm grows, Dennis Stoltzfoos is always thinking of ways to make the farm more efficient. He wants to compost manure that collects near the barn and make his milk parlor portable, which would eliminate the need to bring his cows into the barn from a half-mile away. Portable parlors are not a new idea, other farmers use variations, but his version would be unique. He would model it after a poultry schooner, a chicken house that can be moved across pasture.

"I want to take what we have and put it in a hoop structure, like a 14 by 28 poultry schooner, and scoot it around on the pasture every day," Stoltzfoos said. "Or move the whole thing once a week and never have to bring the cows into the barn. Our portable milking parlor would be very small. That is what I am attracted to. Instead of getting peanuts for your milk, get a good price and have fewer cows. Small is beautiful, and life cannot get any better than on a small family farm."

• Chapter 4 •

Rotational Grazing and Pasture Health

TERMS TO KNOW:

Overgrazing: Allowing animals to graze plants faster than the plants can recover

Vegetative stage: The growth stage in a plant's life cycle before it begins the reproductive stage. In this stage, they are at their most nutritious for cattle to eat.

Dry matter: The amount of feed in plant material after all the water is removed

Riparian area: The transitional areas of land adjacent to natural water sources

Grass-fed cattle farmers need to get the most out of their pastures, and the best way to do this is through rotational grazing. By using rotational grazing, farmers get more grass yield per acre of pasture and can extend their grazing season. In a well-managed grazing system, farmers are even able to support more animals per acre.

The basic principle behind rotational grazing is simple: The pasture is divided into paddocks, and the cattle are left in a paddock until they eat grass down to a predetermined level. Once the grass reaches this level, the cows are moved to the next paddock, where they eat that grass down to a predetermined level before they are moved again, and so on. By the time they make it around to the first paddock, the grass there has had a chance to recover.

Well-managed grazing prevents **overgrazing**, or eating plants faster than they can recover. Grass needs more time to recover when grazed too low. If plants are overgrazed too often, they will die. Poorly managed pastures develop bare spots in the soil, and these bare spots are unprotected from wind and erosion. In an unmanaged grazing system, animals ignore undesirable species of plants, allowing this unwanted species to multiply. The soil also has difficulty absorbing water, and the cycle perpetuates itself. In extreme cases, overgrazing leads to **desertification**, a thinning of plant cover due to weakened soil productivity.

The rotational grazing process mimics the way herds eat in nature. Herds travel packed together in bunches to protect against predators. They eat everything in sight or trample the plants they will not eat, but they never stay in one area too long. By the time they return to their initial grazing spot, the plants the herd left in their wake have recovered. Keeping beef and dairy cattle together forces them to evenly graze each paddock and not just pick their favorite grasses and legumes. The trick to rotational grazing is matching the needs of the cattle with the growth cycles of the grass.

Grass Growth Cycles

Grass begins growing in the spring and starts slowly, but as leaves grow and are able to catch more sunlight, growth accelerates. As daily temperatures rise, growth rate increases for a flush in late spring or early summer. Growth slows during the hottest months of summer and increases again as temperatures cool in the fall. Some grasses go dormant in winter. The remaining grass can be grazed, but it will not re-grow until spring. This is why farmers usually have to supplement their herd's diet with hay or other stored forages in winter. Other factors determine growth rate at specific points during this cycle; for example, a lack of rainfall slows growth.

Most growth takes place in the base of the plant, and when cows overgraze crops, the base is damaged and plants have to tap into the energy reserves in their roots. This energy reserve allows plants to survive winter and grow back strong in the spring. Grazing the plant too low increases the time grass needs for recovery and weakens the root system. If you do not overgraze plants in the growth stage, they will regenerate with no harm done. The remaining leaves will capture sunlight for energy, so the more leaf you leave behind, the quicker re-growth will be. When plants first initiate re-growth, they grow slowly. The amount of rest needed after grazing varies from species to species. At some points, certain plants only need seven days of rest; other plants at different points need more than 30 days of rest.

Plants are most nutritious while they are growing, when they are said to be in **vegetative stage**. In this stage, plants gather energy through photosynthesis and are either storing it or using it for growth. As a plant matures, its nutrient content decreases as it uses its stored energy for seed production. The goal of your grazing plan is for your animals to graze plants before they begin forming a seed head but not so far down they weaken the root system. Rotate your herd back around after the grass has grown back but again before they begin forming a seed head. If your grass is reaching maturity faster than your cattle can keep it grazed, mechanically clip it and harvest it as hay or other stored forage.

Many state extension services, department of agriculture offices, or state Natural Resources Conservation Service offices publish guides with recommended grazing heights for common area plants. These guides tell you what height you should start grazing the plants at, what height you should leave the plants at, and how long the plants need to recover. These guides will also tell you the best time to harvest the plants in your fields for hay. Another good source is the Maryland Cooperative Extension's "Evaluating Hay Quality" fact sheet, which can be viewed at **http://extension.umd.edu/publications/pdfs/fs644.pdf**.

Learn to watch your grass because pastures fluctuate in quality. You can plan when to move your cattle again in part by watching the ground. Keep an eye on the grass in the current paddock to be sure the animals are not overgrazing. If the plants in a paddock are getting down past the recommended level, move your cattle. You can also get an idea of how soon to move the cattle by looking at the paddocks ahead and behind in the rotation, as mentioned in the University of Kentucky fact sheet "Rotational Grazing" (**www.ca.uky.edu/agc/pubs/id/id143/id143.htm**). The one behind will tell you how fast grass is re-growing, and the one ahead will tell you if the grass is ready to be grazed again. If the grass behind seems to be coming in slowly, you might need to slow down your rotation. If the plants in the upcoming paddocks are about to form a seed head, you may need to speed up the rotation. Knowing how to visually gauge the condition of your field will enable you to manage your herd's grazing for optimum quality and yield. This will also help you plan grass finishing because you will know when the most nutritious plants are growing. *You will learn more about grass finishing in Chapter 8.*

Paddocks

Now that you know the principles behind managing grass growth, you need to know how to divide your pasture. This depends on the size of your field and the number of animals you own. This is another time an aerial map of your field will come in handy. Mark all the streams, existing fences, water sites, and hard-to-navigate areas to aid in your planning.

Start with at least five or six paddocks. Generally, the more paddocks you have, the easier it is to manage the plants in your field. If you start with ten paddocks, you could leave your cattle in for three days and then move them. Because of the time it takes to move cattle from one paddock to another, you may want larger paddocks so you can leave your cattle in for more than three days. As you develop your grazing strategy, it might be wise to start with a small number of large paddocks and, if necessary, you can further divide these with temporary fencing. No matter how many paddocks you have, the key is to move the cows before they overgraze an area.

There are many ways to set up a grazing plan. Some farmers with large, flat fields simply divide their fields into 30 or more paddocks and move the herd each day. Many dairy farmers move their cattle after each milking. They often subdivide large paddocks with temporary fences they change for each herd move. Each paddock needs shade and access to water. Some farmers base their rotation around a water source and make a wagon-wheel pattern. With this system, each section of fencing extends outward from a central point like a spoke on a wagon wheel. With this design, you could also think of the field as a pie where each paddock equals a slice. You could also use fencing to make lanes to a centralized water source. You may decide to use more than one portable water tank and design paddocks that share these. You could also arrange paddocks around narrow lanes that lead to available water sources. Try to work with your land's topography. Different areas of the farm require different grazing management strategies. For example, some areas of the pasture grow faster than others, and some areas of your farm, such as those in **riparian areas**, areas adjacent to ponds, streams, and other natural water sources, do not hold up well under herd pressure in wet weather. You may not want your animals there all the time or in wet weather, but you may need to graze these areas occasionally.

Do not worry too much about making mistakes when you first draw up your paddocks; if you use temporary fencing to divide them, you will be able to make adjustments. As you continue to monitor your grazing program, your estimating skills will improve, and you will be able to troubleshoot problems with your system and make corrections.

Water sources

When planning your pasture divisions, make sure you do not force your cattle to walk more than 800 feet to get to a water source. No more than 600 feet is even better. If they have to walk more than 800 feet, they will linger around the water source, tearing up the ground and concentrating manure in a single area. Also, make sure cattle can get a drink when they want. If your source is not large enough, cattle will cluster around the source, and many of them will not be able to get a drink. Cattle need anywhere from 25 to 50 gallons of water each per day. If cattle do not get the water they need, their growth performance will not be at the rate you need it to be.

Natural water sources can be attractive because they are free, but they could dry up at various times of the year or even be dangerous to livestock during freezing weather. If you let your animals use a pond or stream as a water source, you have to set up your fencing to allow only limited access to the water or else they will linger near the water on hot days, which will tear up the banks. Keeping them out of the water will also prevent their urine and manure from polluting the water.

Design the section of fence by a natural water source so they only have enough room to reach in and get a drink, which will prevent them from relieving themselves in the water and will also discourage them from staying too long. If necessary, set up your fencing so there is a gate to a limited area to cross the stream. For a more detailed discussion and pictures of these types of designs, try the ATTRA publication "Paddock Design, Fencing and Water Systems for Controlled Grazing," viewable at **http://attra.ncat.org/attra-pub/paddock.html#water**.

To protect the areas adjacent to the water source, you could devote paddocks to these areas and occasionally let the animals into the paddocks so they can graze the grasses there. However, be careful not to give cattle access when the ground is wet because it will be particularly vulnerable to damage from hoof traffic. Leave the grasses in these paddocks a little higher than in other paddocks, at least 4 inches, to better protect against runoff. In some places in the country, water protection laws restrict cattle access to natural water sources to keep the animals from polluting the water. Check with state environmental or natural resource conservation offices to see what laws apply in your area. If there are regulations, keep the buffer zones and eliminate the access points. Instead, use artificial drinking spots for your cattle.

Water troughs are a good way to ensure clean drinking water for your animals and to prevent pollution in streams and ponds. Portable water troughs are an inexpensive way to ensure your animals can quench their thirst from any paddock. Get portable tanks at farm stores in various sizes, from 15 gallons to more than 900 gallons. You might pay between $100 and $300 a tank. Two companies that ship plastic troughs are Go-To-Tanks (**www.gototanks.com**) and Plastic-Mart (**www.plastic-mart.com**). Some farmers do not like portable water tanks because they must be moved each time you rotate your herd, but this should not be a problem if you rotate your troughs the same time you rotate your paddock. Each has a drain plug you can pull to empty the tank before you move it. You can buy insulated troughs or heaters to prevent them from freezing in winter. The water to fill these tanks can come from many sources, including wells, streams, ponds, or public water systems. Using public water systems provides a reliable source of clean water, but this could be expensive, depending on your local rates.

Be creative and save money on troughs; for example, you could cut a plastic drum in half to make two troughs. Supply water to these tanks with a simple irrigation pipe, available at farm supply stores. Plan your paddocks so several paddocks share pipes. Leave slack in the pipes when you lay them across pasture to give the joints flexibility during freezing and thawing. Treat these pipes the same as you would water pipes or garden hoses around your house to protect from freezing — drain the pipe before freezing so it does not split when ice expands or leave your water source on a slow drip. Bury these pipes if you want, but for a novice farmer, it is probably better to leave them above ground to give you flexibility to make adjustments in your grazing scheme.

There are also different ways to pump water into these systems. Water pumps can be powered by electricity, gas, wind, solar energy, or gravity if the water source is above or below the tank. There are also pumps powered by the cattle's noses. As the animals reach for water, they hit a lever that pumps water into the tank. You can buy full-flow valves, which monitor water levels and automatically refill the tank, for less than $30.

Grazing Strategies

You can use temporary fencing to ration available forage in a process called strip grazing. In **strip grazing**, farmers move a temporary electric fence back a little each day to reveal new forage. As the fence moves back each day, or a couple of times a day in times of fast growth or after dairy milkings, the cattle keep moving forward to the newly available forage. Farmers often use a temporary fence behind the cattle to keep them from re-grazing plants. Farmers can use strip grazing when grass is growing quickly and the cattle need to be moved often, such as on dairy farms. Strip grazing is also beneficial when grass growth is slow and you want to ration your remaining grasses. If the fence is moved back until there is no new forage, farmers can provide supplemental feed such as hay.

If you have grazing corn, strip grazing is also a good way to ration it. But you must be sure none of the stalks come in contact with the fence and short it out. Livestock can knock the corn down into the fence as they move through the paddock. One way to control this is to mow down the stalks near the fence area with an ATV or truck so the cows do not knock stalks into the fence. *Grazing corn was discussed in more detail in Chapter 2.*

Many farmers graze their herds in groups that are each given a different paddock. These paddocks are sometimes separated from each other by wide expanses of pasture. One common reason these groups are separated is because one group has different requirements than another. For example, you might keep a group of lactating cattle on a lusher patch of pasture than stocker cattle not ready to be finished. On a dairy, you might also keep your nurse cows and calves separate from your milk cows. Otherwise calves would leave their nurses and return to their mothers. In seasonal breeding systems, bulls are kept apart from females so they do not start the breeding season too early.

Another strategy is to graze the separate groups in adjoining paddocks. Give the group with higher nutritional needs, such as lactating cattle or beef cattle in the finishing stage, first access to the best grasses. In this case, grass is not grazed down as far as it could be, but the cattle with the greatest nutritional needs can selectively graze what they want before being moved quickly into a new, fresh paddock. After the first group is moved, a new group with lower nutritional needs is brought in to graze the grass down further. This system is called **leader-follower grazing**.

A variation on this is called **creep grazing**. In creep grazing, unweaned calves are allowed to go through an access point in the fencing system so they can graze in the next paddock ahead of their mothers. Even though they are still nursing, calves will begin eating grasses in the field. Usually, a field used for creep grazing is higher quality than the field the rest of the herd will graze. By letting the calves eat the best grasses first, they get the extra nutrition they need from the grass, which also reduces the lactation demands on their mothers.

Creep grazing is accomplished by setting up the fences so there is a small access point calves can get through, but mothers cannot. Mothers may get a little stressed about this arrangement at first, but they quickly realize their calves can return at any time. These access points can be as simple as setting up the temporary fence line high enough calves can go under but the mothers cannot, usually 42 inches. A common way to allow creep grazing is to use creep gates. These can be as simple as opening a regular gate but putting wooden posts in the open area to reduce the size of the passageway. These creep panels are usually about 18 inches wide. Do not confuse this method with the common conventional practice of giving calves access to feeders full of grain, called creep feeding.

Another short-term strategy is called **set stocking**. This is a strategy for animals in the finishing stage (the finishing stage is usually the last couple of months, or last 200 pounds gained, before slaughter). In set stocking, you turn your finishing-quality animals loose in a large, lush pasture and let them eat what they want until they are ready for slaughter. The other animals can be rotationally grazed as usual.

Maintaining Pasture Health

A well-managed herd rotation is going to do wonders for your pasture health. Your soil fertility will improve, your plants will be healthier and hardier, and a good management plan allows the land to provide much more of your animals' nutritional needs. It also reduces the need for supplemental feed such as hay. There will be times, however, when you have to take steps to maintain pasture quality. You have to keep beneficial plants growing and manage the pesky ones. It is important to monitor your soil health by keeping an eye out for weeds by walking your property regularly and by performing soil tests at least every couple of years to ensure it is at its best.

Weeds

Weeds compete with the plants you want in your field for sunlight, water, and soil nutrients. Many types of weeds spread quickly and steal nutrients from other grasses. Many weeds are also unpalatable to cattle. Although cattle can eat Johnson grass, it can become toxic after it begins to wilt. Most of these plants are only fatal to animals if they are eaten in large quantities, but they should be eliminated because cattle will not each as much grass as they need if they keep coming across weeds they find unpleasant. A few weeds scattered throughout your pasture is not a big deal, and many weeds, including Canada thistle and leafy spurge, even contain some nutritional content comparable to common forage grasses in their vegetative states. Weeds become a problem when they gain a competitive advantage and crowd out your desired grass species. Keeping your soil pH in balance and keeping cattle from grazing new plants too soon are steps that will help prevent weed infestation. Weeds gain an advantage when cattle graze grasses lower than the height of weeds. Although there are many weeds cattle do not like to eat, cattle are not as selective when grazed in tight groups because there is increased competition for the available forages.

A weed you may be familiar with is the dandelion. Dandelions are actually fairly nutritious, but too many of them in a pasture is a problem because they spread quickly. Other common weeds are ragweed, quackgrass, and ribwort; these species can come to dominate your pasture and crowd out other types of beneficial grasses that cattle prefer.

You have a few options when dealing with weeds. If your field is infested with tall weeds, mowing the ones your cattle have left will help the grass in your pasture regain control. Be sure to mow them before they form seeds — after the seeds are formed, they can spread in your field even if the plant is cut. Herbicides also work, but you have to observe a waiting period before you can return your animals to treated areas. Each product should tell you how long to wait before it is safe to graze again. Some waiting periods are as short as three days or fewer, and some as long as 30 days or more. Common synthetic herbicides include Roundup, which kills everything, including surrounding grasses; 2,4-d, which will kill broadleaf plants such as dandelions; and Crossbow, which targets brush and woody plants such as poison oak.

Synthetic herbicides are not an option for farmers aiming for organic certification. Instead, vinegar is a common ingredient in organic herbicides. Use a blowtorch to burn brush or other plants you do not want growing in your pasture, but do not use one if your field is dry from lack of rain because dry fields are more likely to catch fire. Another method farmers use is to graze cattle in tandem with other species of animals. Goats and sheep, for example, eat weeds cattle will not.

Planting in the fall gives your plants a chance to establish themselves before the spring, when plants, including weeds, grow quickest. Be sure to read the label on seed mixtures. The label usually indicates the likelihood of weed content. Mob grazing, or grazing a large number of animals in a small area for a short period of time, is an effective way to eliminate many weeds because a large number of animals in a small space will trample most everything they do not eat.

Regular soil tests

It is also important to monitor your soil health. Conduct soil tests every two or three years to ensure all the nutrients are in balance. You can test more often if you want. Correcting soil imbalances helps keep weeds away because each weed appears in response to a specific condition in the soil. Keeping your nutrients balanced and your soil fertility at its best levels will ensure your desired forages can thrive. Follow the recommendations provided by the lab that analyzes your sample. You may also have to add fertilizers or nutrients each yea. This continuing care should be recommended on your soil test.

Annual seeding

If your grazing strategy depends on annuals, you will have to plant these grasses every year. Annual seeding is accomplished in the same ways as the plantings, by using a broadcast seeder or no-till drill. The best times to plant are spring and late fall. Local universities and state Natural Resource Conservation Service offices publish plant guides that tell you the best plants in your area and the best times to plant. Keep your cattle out of paddocks in which you plant so the animals do not harm the newly sprouted plants before they have a chance to establish themselves. The NRCS planting guides will tell you when it is safe to let your cattle graze.

Keeping records

If you want to be precise, keep records of pasture rotations to better gauge pasture performance. Write down the dates animals were in a specific paddock, how long they were there, how tall the grass was when they got there, and how tall the grass was when the animals moved out. Check on each paddock and write down how much the grass grows back in a week, ten days, two weeks, and so on. Recording this type of information tells you precisely how much time each paddock needs to grow back and allows you to make adjustments to your rotation in the future.

Detailed records over time will also let you see how your grazing management improves pasture quality and available grazing days. *The next chapter discusses how to acquire the animals you need to harvest your pasture grasses.*

CASE STUDY: HOBO RANCH

Dan and Mary Flitner, managers
Hobo Ranch
Las Vegas, New Mexico
flitner@plateautel.net
http://hoboranches.com

Dan and Mary Flitner's standing offer is this: Anybody, from anywhere, anytime, can come take a free tour of Hobo Ranch. They rarely get any takers because nobody lives close enough. The farm is 35 miles east of Las Vegas, New Mexico, in the state's eastern prairies. They are in between a few rivers but not close enough to irrigate. Their main ranch is 16,000 acres, but much of it is cliffs, rocks, and canyons that will not support many cattle. But it will tear up your pickup if you drive more than 3 miles an hour.

Hobo Ranch might be one of the most difficult places in the country to pasture finish beef cattle. The growing season is short, from May to October, and centuries of overgrazing robbed the land of much of its cool-season grasses. The average rainfall is 12 inches per year. The rules of thumb are you need about 35 acres of your best land to support one animal, and on the worst, about 150 acres. But since the Flitners began managing the land in 2004, the fields have continued to improve.

Back then, the Flitners were looking for a lifestyle change. Dan Flitner grew up in conventional agriculture; his family ran a feedlot and a farm. He said they ran themselves ragged trying to keep things going and never seemed to get paid enough for the amount of work they were doing. Then, he, his father, and his brother all had health scares. He does not believe the problems were caused by the beef they were eating, but he wanted to make a change. When Mary Flitner's parents offered to let them manage some land they had bought in New Mexico, the couple jumped at the chance.

Dan Flitner is not sure his grass-fed cattle farm is any less work than a feedlot, but the margins are better. He sells his grass-fed beef for $7 a pound. He finished 25 steers last year. As the operation continues to expand, he may finish as many as 100. He started his herd with 20 head of cattle bought in 2006 from a farmer in Colorado who raises grass-fed cattle. Now, Flitner's herd includes almost 100 females, 35 of them purchased, the rest born on the farm. He bought 160 steers from neighbors that he hopes to finish on grass. His animals spend their whole lives in pasture. "They have been out their whole lives when I get them, and they stay out," he said.

He let his field lay fallow for two years before using it for grazing. He did not plant anything, but his rotational grazing program is already paying dividends — cool-season grasses are starting to return. Still, without irrigation, his family's land is not good enough quality for finishing, so he rents two other plots of irrigated pasture. One is 200 acres that he splits into three to five paddocks for 30 head of cattle. Last year he began leasing another 600 acres a little bit north of the town. He planned to split that into ten or 15 paddocks for 180 head.

During the dormant season, he moves the cattle about once every two weeks and feeds them alfalfa hay to supplement. He tries to move them a minimum of once a week beginning in May. He would like to move them more often, but it is difficult to move fences and water supplies that often on so many acres. He strings a mile of polypipe to various water sources, and he tries to plan his paddocks so he can stay a month at each watering hole.

He finishes his Angus crosses at between 18 and 24 months of age. On the rental pasture in spring, especially after a good monsoon season, the animals will gain 2 or 3 pounds a day. But winter is tougher — the cattle only gain about ½ pound a day. This winter, Flitner is planning to lease irrigated pasture just a bit further south near the Rio Grande where he could pick up an extra 10 or 15 degrees in winter that would provide more grass and possibly help his animals gain 1.5 pounds or more per day. This would help finish his best animals quicker and give him more time to finish his slower-gaining animals. His animals will finish at between 950 and 1,200 pounds. He judges this based on each animal's frame type and how it fills out. "You can tell when they are starting to get a little overweight, and that is a good thing when they do," Flitner said.

In an area as unpredictable as his, he tries to plan several months ahead and makes sure he always has enough hay reserved or extra pasture to fall back on if there is a drought or other emergency. Flitner usually will not need his hay until the end of the year; in a good year, he can go until March. This was a tough winter, and he had to start feeding earlier than he wanted to.

"Since the end of January, we have had 4 feet of snow this year," Flitner said. "When it started snowing like that, I doubled my feed rate from 5 pounds to 10 overnight, which still is not a lot of hay if you are making it instead of buying. It drives me crazy, but it will be worth it. Our cattle look good, and the range is going to look good when this starts to warm up and we start to benefit from all that moisture. If we get a normal monsoon this year, we will maintain good gains all summer."

• Chapter 5 •

Building Your Herd

TERMS TO KNOW:

Polled: Having no horns; usually naturally hornless

Bull: An uncastrated male. Males that are not castrated are used for mating with females.

Steer: A castrated male

Heifer: A cow older than 1 year old that has not had a calf.

First-calf heifers: A cow that produced just one calf

Weanling: An animal that has just been weaned

Yearling: An animal between 1 and 2 years old

Artificial insemination (AI): The process of breeding cattle by using harvested semen rather than mating a cow with a live bull

Custom livestock haulers: Contractors who will transport animals to or from your farm for a fee

Biosecurity: Procedures to keep new illnesses from infecting your animals

Herd sire: The bull that contributes most of the genetics to the calves born on your farm

You need the right animals for your grass-based farm. You want animals suited to your climate, that can grow strong and stay healthy on a pasture-based diet, and that produce good quality meat or milk. You have many options when it comes to breeds. There are dozens of breeds in all shapes, sizes, and colors. Each breed has characteristics desirable to certain farmers. For example, the Angus breed has a reputation for producing high-quality meat.

Farmers who convert from conventional farming operations to a grass-finishing farm usually work with the animals they already have. If you are considering such a switch, you know your animals are adapted to your climate. You will have to monitor the cattle that do not do well with the lifestyle change and sell them. For example, the animals that do not gain weight as well as they might on an all-grain diet or stay in good condition during milking without grain supplements should be sold. Your herd genetics will improve as you mate the heifers that adapt best to the pasture with good-quality bulls. You can also breed the animals on your farm with breeds strong in characteristics you want to improve in your herd, such as breeding with Angus cattle to improve meat quality or breeding with Brahman to improve heat resistance. *Breeding will be discussed later in this chapter.*

If you are starting a herd from scratch, choose what breeds you want to work with. Although certain breeds are used more often than others on grass-fed cattle farms, many farmers say breed is not as important as choosing the best individuals from within a breed. The cattle used most often in grazing systems are smaller with wide bodies because it is easier to pasture finish a 1,100-pound animal than one that needs to grow to 1,500 pounds. Dual-purpose breeds used for both milk and meat are often used in pasture-based systems because of the flexibility they provide in the types of products you can sell. On most dairies, male calves are not desirable because they cannot be milked, but if you use a dual-purpose breed, the male calves born on dairy farms can grow into acceptable beef animals.

If you are buying animals, you also have flexibility in the ages of animals you buy. Buy cows with newborn calves; or buy weaned calves, which have been taken off milk; or yearlings, between 1 and 2 years old. Keep these animals just through the growing season and then sell them, or keep them until they reach their ideal weight for slaughter.

Appearance of a Healthy Animal

Farmers are always looking for certain traits when purchasing cattle. You want a masculine-looking bull with large, rounded shoulders and a big scrotum; these features signify appropriate testosterone levels and virility. He also needs to have no injuries or problems with his legs so he can safely mount females. You want fertile cows with feminine traits, such as wide, calf-bearing hips. You want your breeding stock to produce calves that can grow quickly and mature early but are not born so heavy that birth is difficult on the mother. These types of traits can be predicted based on the animal's own growth statistics and the statistics of all the calves fathered by its parent. After you have farmed for a while, you will be able to see which mothers produce the best or fastest-growing calves.

At an even more basic level, you want animals that look healthy. A healthy cow has clear eyes, erect ears, a healthy coat, clean muzzle and nostrils, and a good appetite. Consider having a veterinarian check out animals before you buy them from a private seller. You or a vet can take an animal's temperature with a rectal thermometer. His or her temperature should be around 101.5 degrees. Animals should be able to move effortlessly and without pain. If they favor a leg or seem to have pain when they move, they could have an injury or illness you do not want to deal with. If you are buying animals, stay away from the sick ones because you do not want to bring illnesses onto your farm. Look for reputable sellers who vaccinate their animals. They should be able to show you records of what they vaccinated for and when the vaccinations were given.

Breeds of Beef Cattle and Their Qualities

Most breeds have breed associations that describe the pros, cons, and histories of the animals. Oklahoma State University also has an excellent database of cattle breeds and qualities at **www.ansi.okstate.edu/breeds/cattle**. Another good source of information is the American Livestock Breeds Conservancy, a nonprofit membership organization that works to protect rare livestock breeds from extinction (**www.albc-usa.org**). Talk to farmers in your area to see which ones are popular and successful in your region. Many farmers use crossbreeding programs, which means they mate cattle from one breed with cattle from another breed to get the best qualities of both breeds. *Crossbreeding is discussed in more detail later in this chapter.* To get you started, here is a list of breeds and some of their qualities:

Angus

Angus cattle are a favorite breed in beef production because their carcass yields well-marbled meat, which means fat fills in between the muscle tissues and looks almost like a pattern you might see in marble. Marbling contributes to juiciness and taste. This Scottish breed is solid black or red and naturally polled, meaning they are born without horns. Angus cattle are found all over the country and can do well in harsh winter conditions. Angus calves are born small, which means easier births, but they grow quickly and reach weights comparable to other breeds by the time they are weaned.

Brahman

The hump on a Brahman's back and its long, floppy ears make the breed easily identifiable. They are usually red or gray. The Brahman is an Indian breed that tolerates hot weather and insects. This makes it an important animal in crossbreeding programs in warm parts of the country. They are known for having longer reproductive lives than many breeds. Brahman yield leaner meat.

Hereford

Like the Angus breed, Hereford cattle are highly desirable for beef production. They are easily recognizable with white faces, white tips on their tails, and red bodies. Most have thick, curved horns, but one strain is naturally polled. They are a British breed found in most parts of the world. One problem of the breed is it is prone to cancer eye because it does not have pigmentation around its eyes to protect it from the sun. If eye cancer is not caught early, it can become extremely uncomfortable for a cow, and you will have to slaughter it or sell it at auction to someone who will slaughter it.

Galloway

The Galloway is a polled breed with good-quality meat that marbles well. They are known for fattening up on low-quality forages, according to the American Galloway Breeders Association, which makes them efficient grazers. The breed originated in Scotland, but it is also considered a British breed. Galloways are usually black animals with long, thick hair that enables them to withstand harsh winter conditions. Some are light brown or yellow-gray. The Belted Galloway is an easily recognizable creature, with a white stripe around the back of its midsection. There is also a rare breed of White Galloway.

Charolais

Charolais cattle are popular in crossbreeding programs, when you use purebred bulls of one breed with cows of another breed, because they produce large amounts of high quality meat. They mature later than other breeds, which means they fatten to heavier weights. They are heavier than what many grass-fed producers target, but there are grass-fed producers who have success with Charolais cattle. This breed is known for its performance in warm weather, but it can also tolerate cold. Charolais cattle are a French breed, and they are white or cream-colored.

Limousin

Limousin cows have high yielding, muscular carcasses, but there is less marbling in their meat than in other breeds. Limousin cattle can be golden-red or black. They are a French breed known for having a more unpredictable temperament than most breeds, but the North American Limousin Foundation aggressively targeted docility in its breeding programs.

Chianina

Chianina cattle put on weight quickly, with some farmers reporting Chianina crossbred calves weighing dozens more pounds at weaning than other types of calves on the farm. They are often used in crossbreeding programs to hasten growth in calves. This Italian breed is white with black on the tail. Chianina cattle in the United States are among the largest in the world. They are tall, muscular animals with long legs.

Simmental

The Simmental breed is known for producing large amounts of high-quality beef and has a heavily muscled back and loins. It is another breed born small but with quick growth. Simmental hair can be yellow, gold, or dark red, with white hair on the head and lower legs. Some animals are also black. This Swiss breed has a reputation for being docile.

Scottish Highland

This is a durable breed that can survive the harshest winters and perform on low-quality pastures. Their heavy outer coat can be various colors, including red, yellow, or black. Highlands produce tender beef that is much leaner than the meat from most breeds because their thick coats keep them warm and reduce the need for fat.

Breeds of Dairy Cattle and Their Qualities

Grass-fed dairy production is different than conventional dairy production. Conventional dairy farms breed their animals to produce unnaturally large quantities of milk. Since 1970, average milk production per cow has nearly doubled, from 9,700 pounds per year to nearly 19,000 pounds, according to the USDA. Milk is often measured in pounds, not gallons, and 1 gallon equals about 8 pounds. This increased production is made possible by feeding the cows grain, which does not have as many vitamins and nutrients as pasture food sources. Milk from grain-fed animals does not have the nutritional quality of grass-fed dairy animals because milk from grass-fed cattle is higher in good fats, vitamins, and minerals. This large quantity of milk works against farmers when they sell their milk as a commodity. The price farmers are paid for their raw milk has been in a prolonged slump; it dropped anywhere from 40 to 50 percent from 2008 to 2009. The low prices have made it less profitable to produce milk for mass markets and have forced many dairies out of business. The good news is dairy products targeted for niche markets, such as for customers who want grass-fed or organic products, can draw higher prices than milk sold at supermarkets.

Good dairy milk is high in protein, butterfat, and lactose and has a low somatic cell count. High levels of somatic cells are considered an indication of mastitits, or an inflammation of the udder. Dairy processors set limits on the somatic cell content they will accept, and they pay more for low counts. These limits vary by state. The limit set by the federal government is 750,000 cells per milliliter. Measure this yourself with a California mastitis test (CMT) kit, which you can buy at farm supply stores or at livestock veterinarian offices. Whole milk is served with all the fat intact; 1 percent, 2 percent, and skim milk are sold with varying amounts of the fat removed. The percentage of fat in milk varies by breed. Some breeds produce milk with extra butterfat, which makes it creamier and useful for making products such as cheese and butter.

The five most popular dairy breeds in the United States are Holstein, Jersey, Guernsey, Brown Swiss, and Ayrshire. Some of these breeds work well in grazing-based systems, while other breeds are not as successful. Here is a look at a few dairy breeds:

Holstein

Holsteins are the most popular breed in conventional dairy systems because of their incredible milk production. The best producers can provide 25,000 pounds of milk per year. Holsteins are black and white or red and white, and the breed was developed in the Netherlands. They can weigh more than 90 pounds at birth and grow to about 1,500 pounds at maturity. Their reputation in grass-based systems is not as good because the most common genetic lines were bred to produce lots of milk in grain-based, confinement dairies. They do not do well on all-grass diets.

Jersey

Jerseys are a small-sized, popular breed of dairy cattle that often do well on grass-fed cattle farms. Many rural families kept a family cow for milk until about 1970, and Jerseys were the most common choice because of their easy-going disposition and high-quality milk. Jersey milk is higher in butterfat and protein. Jerseys are docile and are a range of brown colors, from copper to dark brown. The breed developed on the island of Jersey in the British channel. A mature Jersey weighs around 1,000 pounds or fewer. Their calves are also smaller at birth.

Brown Swiss

The Brown Swiss is another large breed, but this breed originated in the rugged Swiss Alps and proves to be more successful in pasture-based dairy systems than Holsteins. These brown cows weigh 1,500 pounds at maturity. They do well in all weather conditions and are noted for their ability to withstand extremely warm temperatures. They produce milk rich in butterfat and protein. They are easy to care for and laid back.

Ayrshire

Ayrshire cows have a reputation for pasture-grazing success. They can thrive in all climate conditions, from Africa to Scandinavia, according to the Ayrshires Cattle Society. The average cow from this red and white Scottish breed will weigh around 1,200 pounds. The calving process is usually smooth, and calves are usually strong and healthy. They are a horned breed but are usually dehorned at birth.

Guernsey

Guernseys are medium-sized cattle with a mature cow weighing about 1,100 pounds. Like Jerseys, Guernseys were also a common choice for rural families that kept family milk cows. They produce golden-colored milk because it is high in beta carotene, a source of vitamin A. The milk is also high in butterfat and protein. The breed developed off the coast of France, and its colors are fawn and white.

Dutch Belted

This is a medium-sized breed that produces milk billed as ideal for drinking because of its high butterfat content. They are usually black with a white belt between the shoulder and the hips. The breed is so rare the Dutch Belted Cattle Association of America does not recommend breeding Dutch Belted females with other breeds. Rather than cross breeding, the association recommends all calves produced by Belted cows be purebred Dutch Belted. However, Dutch Belted semen can be used on cows from other breeds.

Montbeliarde

Montbeliarde cattle are another French breed that produce high-butterfat milk ideal for making cheese. These animals, which have red and white patches, are known for fertility, reproductive longevity, and their penchant for giving birth easily and with no complications. They also have low incidences of mastitis, a common udder disease. *Mastitis will be discussed in more detail in Chapter 6.*

Dual-Purpose Breeds

Although most cattle are bred to excel at either beef or dairy production, a handful of breeds are known to work well in both systems. These animals present an opportunity for dairy farmers who want to diversify their income. Here are a few breeds with a reputation for dual-purpose production.

Devon

This type of cattle is a small-framed, docile animal that does well on grass. It is hardy in winter and thrives on low-quality forage. Members of this English breed were used as oxen in Colonial America. They are known for producing sweet meat. A close relative is the Milking Devon. Both the standard Devon and Milking Devon can be used for meat and milk.

Shorthorn

The shorthorn is an adaptable English breed that yields good-quality meat and produces a high volume of milk. The aptly named shorthorn is typically red, white, or roan, a mixture of red and white. Separate breeding lines, the shorthorn and the milking shorthorn, were developed so each could specialize in either beef or dairy. Many producers still use these lines for their specialty, but both lines are used for the dual purposes of meat and milk.

Criollo

This small, horned breed produces tender, lean meat. It is a Spanish breed that does well in heat but is adaptable to most environments. It is known for eating less-than-ideal forages. Criollo came to the United States on Christopher Columbus' second voyage to America in 1493, according to the American Criollo Beef Association. Criollos thrive in arid areas and are efficient with available forage. Criollos are kept as dairy animals and provide high-fat milk.

Normande

This French breed developed on the pastures of Normandy, France, and produces milk ideal for cheese making. Their carcasses are considered excellent for beef. These cows have small heads on medium-sized bodies. They are adaptable and sturdy, and they are billed as "the ultimate grazer" by the North American Normande Association because they are built to navigate rough terrain and can grow well on less-than-ideal pastures. Their coats feature patches of various sizes, and their colors range from mostly white to fawn to red-brown. Rings around their eyes provide protection from the sun. They reproduce well, and calves grow quickly.

Dexter

Dexter is the smallest cattle breed in North America, with bulls reaching just 46 inches in height, according to the Purebred Dexter Cattle Association of North America. A full-grown Dexter cow will weigh fewer than 800 pounds; a bull might weigh fewer than 1,000. They are usually black, but they can be red or grayish-brown. They can be polled or horned. This Irish breed does well outdoors year-round in all climates. Dexters produce lean meat and smaller cuts than larger breeds. Their milk is high in cream and butterfat content.

Where to Find Animals

Before you buy cattle, get an idea of the fair-market value of the animals you are planning to purchase by checking prices in newspapers or online or by talking to experts you trust. The USDA has a website (**http://marketnews.usda.gov**) dedicated to market reports that shows up-to-date prices farmers are getting for their cattle. This will vary by breed and by current market conditions. If you have a fair price in mind before you purchase your cow, it is less likely someone will take advantage of you. As you get more experience raising cattle, you will probably develop more ideas of when you might be able to find a bargain. For example, prices for calves are usually lower in the fall than they are in the spring because this is when most farmers sell newly weaned calves.

Buy from farmers who raise their animals in conditions similar to yours. These farmers should raise their animals to your standards — you want animals who have been outside their whole lives and who have not been introduced to grain.

Private sales

Private sales might be the safest bet for a first-time farmer. Breeders selling bulls usually advertise in the classified sections of local newspapers or in breed association newsletters, which can be ordered directly from the associations. If you buy a bull, associations are often notified by the breeder and will begin mailing you literature. You can also find breeders through Internet searches for breed associations. For example, the American Angus Association is at **www.angus.org**. You may also receive suggestions by asking knowledgeable people in your area, such as the staff at your local feed mill.

If you are looking for a specific breed, most of them have national or state associations that help link sellers with buyers. These can usually be found with quick online searches for a specific breed and its association. For example, the North American Limousin Foundation (**www.nalf.org**) includes links to state associations. If you target a popular breed, you may be able to find a neighbor nearby who has animals for sale. This way you will be able to tell what conditions the animals are raised in and if they would do well in your system.

Public auctions

One place to buy and sell cattle is sales barns, also called stockyards, where farmers gather for regular public auctions. These auctions include hundreds of animals from dozens of sellers. Most stockyards have sales about once a week. Before you go to an auction to buy an animal, it would be a good idea to go and observe so you can see the process first and know what to expect. You will also be able to get an idea of how your judgment of animals compares to that of other buyers.

Sales barns may not be the best places for a new farmer to build a herd because these auctions often include many **cull animals**, animals farmers decide to get rid of for various reasons, such as infertility or susceptibility to illness or parasites. You will not know who you are buying from, and you may not have a way to verify if the animals are illness-free or if they are pregnant or ready to be bred. Sales barns also carry some risk because animals are exposed to so many other animals that could potentially transfer illnesses. Exceptions to this would be when stockyards sponsor breeder auctions or auctions from other certified programs. **Breeder auctions** are held for animals with certain genetics, such as a Hereford breeder auction. These are held at stockyards or at another site. Sometimes, auctions can also be held for estate sales or for farms going out of business.

If you decide to buy at an auction, one of the goals is to stay within budget. Arrive with a maximum spending amount in mind and stick to it. Also, be sure you know how you will transport animals after they are purchased. If you do not have a cattle trailer, perhaps you could borrow one from a friend. If you call the sale barn beforehand, they could recommend custom livestock haulers who will transport your animals to or from your farm for a fee.

Online auctions

Online auctions are popular because animal sellers can show off their sale animals without taking them off the farm, which reduces stress and potential exposure to illness in the animals. Similarly, buyers like these forms of auctions because they do not have to physically attend an auction, and they can make their choices and hire someone to haul the animals to them. These auctions expose sale animals to a wide range of potential customers. Breeder auctions are often held online or via satellite.

Most people involved with online auctions will be honest, but there are risks involved. Protect yourself by getting all the terms of your deal in writing and making sure you know all the details of the deal. For example, will the seller ship animals to you or hire a third party to do this, or are you responsible for picking them up? Sometimes, you can provide a down payment (sometimes by overnight check or wired funds) and will not have to pay the total amount until the animals arrive at your farm. After you buy animals, it may be a good idea to go visually inspect them before they are shipped or make it clear under the terms of the sale you will return any unacceptable animals at the seller's expense. Good auction sites want to ensure their buyers and sellers are protected, and you can contact these companies to see what protective measures they recommend.

Breeding Animals

Dairies and cow-calf beef operations depend on the breeding process to make money. Cow-calf beef operations need a new supply of animals each year that will grow on the farm until they are sold later. Dairies need cows to reproduce so they can be milked. When you breed cows, plan to breed them when they are 14 to 15 months old so they give birth around the time they are 2 years old. When they are bred, they should be about 65 to 70 percent of their mature weight. For example, if a cow's expected mature weight is 1,000 pounds, she should weigh at least 650 pounds when bred (estimate an animal's mature weight based on breed and frame size. Animals with larger frames will be heavier than smaller-framed animals).

You have to be careful about exposing cows to bulls until they are more mature. Heifers that breed much earlier than 14 months often have calving problems, including calves that are not in the correct position at birth. *This problem will be discussed more in Chapter 6.* If you breed your own heifers and want to keep some for breeding to replace your older or less fertile cows, select the ones heaviest at weaning because these are the ones born first, which indicates high fertility, and grew the fastest.

When choosing which females to breed, check pelvic measurements to determine how easy the birthing process will be. A good cattle breeder who is selling breeding stock would have those measurements, which indicate if the birth canal will be wide enough for an easy birth for her expected offspring. If you want to find out about a heifer born on your farm, the best time to measure the canal is three weeks to a month before breeding. A veterinarian can do this for you by measuring the height and width to determine how many square centimeters the birth canal is. For a 600-pound heifer expected to grow to 1,100 pounds or so, an average measurement is 140 cm squared. The pelvis should be taller than it is wide, at least 12 centimeters high and 11 centimeters wide, according to the University of Nebraska paper "Pelvic Measurements for Reducing Calving Difficulty." You might not want to breed heifers with birth canals smaller than this because they are more likely to have calving difficulties.

If your farm operation includes breeding cattle, either use a bull to do it naturally or use artificial insemination (AI). Bulls are expensive and high maintenance, and even if they seem tame, there is always a chance they could attack humans. If you plan to calve seasonally, you must separate the bull from cows able to re-breed or from heifers able to breed because the bull could impregnate your animals outside of your desired breeding window. Or share or lease a bull from a farmer who uses a different calving season than you.

Natural breeding

Find bull sellers the same way you find other cattle: in trade publications, classified advertisements, online, or through mutual connections. If you use a bull, get a veterinarian to examine him before you purchase him, and you also need your vet to examine him on an annual basis. He should be healthy and free from disease. His penis and sheath should not be deformed. The vet will measure his scrotal circumference — usually, the larger the scrotum, the higher the sperm count. Just to be sure, a sperm sample will be collected. His sperm sample should have a minimum of 70 percent normal cells. One measure of a satisfactory bull is sperm motility, or percentage of sperm that move forward aggressively. Acceptable sperm have a minimum motility of 30 percent. Very good sperm will swirl rapidly; sperm considered poor show sporadic oscillation, meaning they only swirl at uneven intervals. The bull's breeding potential may be marked down if the sperm sample shows other cell abnormalities, such as being oddly shaped.

After this exam, the veterinarian will rate the bull based on his breeding potential. Bulls are categorized as:

- Satisfactory, which means they meet all minimum values and are free from defects and infections.
- Unsatisfactory, which means they do not meet minimum values and have defects that will impact breeding potential.
- Classification deferred, which means the bull cannot be excluded from the satisfactory category, but they need to be retested to see if they improve after time or therapy. An example of a problem that would put a bull in this category is an injury that may heal in the future.

A good bull will have a high libido and be eager to mate, but keep in mind a high libido does not guarantee fertility. A simple way to test libido includes putting a bull in a pen with a female ready to breed. If he successfully mates within five minutes, he probably has a good libido. If he shows no interest, this is probably not a good sign. Mature bulls can service more cows in a breeding season than young bulls — ten to 20 cows for yearlings, about 30 cows for a mature bull, and sometimes many more for the best bulls. A yearling also will not have been used yet for breeding, so you will not be able to see how successful he has been in a breeding program. There will not be any data about calves he has produced.

Disposition is important when choosing a bull. If you walk up to the fence to look at him, he should stay calm. He can show signs of being on alert, such as ears standing to attention, but he should not bolt away. He should not move toward you as if to attack. If he runs away or runs at you, you do not want him. If you have a large herd, you may need more than one bull. If you decide to use more than one bull, one bull should be the dominant bull that will sire most of the calves in the herd. The animals will have to sort out their place in the hierarchy themselves, so keep your bulls together in a paddock away from the females for about 30 days so they can establish which one is dominant. If they do not establish their order before the breeding season, some of them could attack and injure the other bulls while they are trying to mate. But if they establish their place, the other bulls can step in and do the breeding at times when the dominant bull is tired. When using two or more bulls, those about the same size and breed and that are raised together seem to do the best.

Artificial insemination

Artificial insemination is a common choice for people who do not want to add a bull to their herd or who want to use genes from top-quality bulls. AI may be a cheaper alternative than caring for and feeding a live bull. Semen comes in containers called straws that often sell for $20 or less, though semen from the most in-demand bulls can cost more. AI may also be more affordable if you are looking for traits from an expensive breed or rare breed. In the case of some breeds where bulls are not available for purchase, AI may be your only option.

The downside of AI is it is more labor intensive because you have to be sure your cows and heifers are in heat so you can inseminate them at the right time. They must be observed for at least 30 minutes twice a day to detect heat. This usually means watching them at morning and at night. If you notice signs of heat in a cow, breed her within 12 hours. Artificially inseminating a cow takes some training and practice, and many farmers prefer to pay an AI technician to perform the actual breeding.

The most reliable sign of heat is called **standing heat**, which means she allows other cattle, including other females, to mount her as she stands. Secondary signs of heat include pacing or restlessness. Cows in heat may group together, follow other cows, and sniff, nuzzle, or lick the rear ends of other cows. A female that mounts other females may be in heat; this is called **riding**. Rough or rubbed-off hair on the base of the tail can indicate a cow has been ridden. Another possible sign of heat is a string of clear mucus hanging from the vulva or smeared on the hind legs or tail. Bloody mucus can appear two to four days after a cow was in heat; if you see this, watch her for her next heat cycle in 15 to 21 days.

Perform the pregnancy tests shortly after an attempted breeding. A veterinarian tests for pregnancy by performing a rectal palpation of the cow's uterus and ovaries. Dairy cattle are generally checked for pregnancy between 28 and 35 days after being bred. Cows that are not pregnant can be re-bred. Pregnancy testing also helps you learn an expected due date. Another indication of pregnancy is failure to return to heat when a cow's next heat cycle is due. Non-pregnant cows are called **open cows**. Some farmers also use what is known as a clean-up bull, whose job is to naturally impregnate cows that did not conceive with AI.

Heredity and crossbreeding

Many farmers turn to the practice of crossbreeding to improve their herd performance. **Crossbreeding** is mating cows with bulls of different breeds. For example, you might mate a cow from a high-yielding breed with a bull known for its fertility or for its resistance to disease. Many farmers in the south use Brahman genes so their herds can tolerate the heat. Farmers in the north may prefer breeding the Galloways for their thick coat. Farmers who try to produce both milk and meat have success breeding traditional dairy breeds with traditional beef breeds. For example, you could cross an Angus bull with a Jersey cow. In crossbreeding programs, the best traits of each breed often surface in the offspring. A weakness in one breed can be offset by mating it to a breed strong in that trait. For example, many farmers will use a Brahman bull to increase heat and insect tolerance.

Mating two purebred animals can result in healthy, high-performing offspring by bringing out the best from both breeds. Bringing out the best in both breeds is called **heterosis**, also known as **hybrid vigor**. It is worth noting that purebred livestock does not mean registered purebred livestock. Registered purebreds are bred for reasons other than performance on the farm, such as lineage or because they look good in a show ring.

Should you decide to crossbreed your cattle, there are several strategies you can implement.

- **Crossbreeding using two breeds:** Cross two breeds and then mate the resulting heifers with a purebred bull of one of the two original breeds.
- **Crossbreeding with three breeds**: Cross two purebred breeds and then mate the resulting heifers with a third breed. Some experts say combining three breeds can bring out the best traits in all three breeds. For example, animals sired by Holstein bulls would always be bred to Jersey bulls, and then those mixed offspring would be bred to a Guernsey. The University of Kentucky recommends using colored ear tags to identify the offspring of each sire. Use one color for the Holstein's offspring, another color for the Guernsey's, and another for the Jersey's.
- **Terminal crosses**: A terminal cross is the offspring of two purebreds that will be used for beef but will not be re-bred. The reason for terminal crosses is the offspring of crossbred animals can be unpredictable. If you breed two crossbred animals, the calves could come out looking like purebreds of either breed, or a some kind of mixture. There will be no way to know beforehand.
- **Buying off-farm replacement heifers:** Although many farmers use heifers born on their farms for breeding stock, you could give yourself genetic flexibility by buying replacement heifers from someone else's farm. This allows you to introduce traits from breeds not found on your farm. For example, you could buy crossbred replacement heifers to breed with your purebred bull. Or, if you only keep one bull, you could use him for three or four years and then switch to a sire of a different breed. AI gives you flexibility in switching sire breeds.

EPD — Expected Progeny Differences

Genetics play a large part in the development of each animal and its usefulness in a beef or dairy production system. For example, genetics are a big influence on meat tenderness. Although many different breeds can do well on grass-finishing systems, individual animals vary in their growth potential. An animal that is 600 pounds at weaning probably will be easier to finish than an animal that is 400 pounds at weaning. Selecting the right bull is critical because genetics are big determinants in growth potential and meat tenderness. If you use only one bull, his traits will be passed on to all the calves born on your farm. If you choose your bull poorly, such as a not-very-fertile one or one that produces substandard offspring, it can be crippling to your farm's earning potential.

One tool many farmers use to select their **herd sire**, the bull that contributes most of the genetics to the calves born on a farm, is called Expected Progeny Differences (EPDs). EPDs are scores based on individual traits that give you an idea of how future offspring, or **progeny**, are expected to perform compared to the offspring of other animals. EPD scores are based on the performance of the animal's parents and on the performance of the individual animal. Breeders from around the country report this data to breed associations that maintain a database so buyers can make comparisons of potential bulls. For example, farmers would weigh a particular bull at birth, at weaning, and at 1 year old and report these scores to the association, which would then calculate the bull's EPD scores against the expected performance of other bulls in the breed.

These scores are useful in comparing animals within a specific breed. Some experienced farmers have trouble reading EPD scores and do not use them to make decisions, but others do depend on them to choose the bulls that will influence the makeup of their herds. Breeders will provide you the scores for their bulls, shown as comparisons to the breed average. For example, if an animal's birth weight score is +10, that means its offspring are expected to weigh 10 pounds more than the breed average. For an animal whose offspring perform at the breed average, its birth weight score would be 0.0. Look at the scores for one or more bulls and decide which one is most likely to produce the types of calves you want.

Common categories EPDs predict include growth, maternal, carcass, and ultrasound. Each is discussed below.

- **Growth**: The growth category predicts the growth potential of a bull's calves. Traits measured are:

  Calving ease direct (CED) predicts the likelihood a bull's calves will be born without difficulties if he is bred to a first-calf heifer. A CED score is a percentage of unassisted births cows are likely to have when delivering a particular bull's calf. An unassisted birth is one in which a farmer does not have to help pull the calf out. A higher value means first-calf heifers are more likely to deliver a bull's calf without trouble. **Birth weight (BW)** predicts the pounds the bull's calves will weigh at birth. **Weaning weight (WW)** predicts how big a bull's calf will grow before it is weaned. **Yearling weight (YW)** predicts how heavy a calf will be at 1 year of age. **Yearling height (YH)** predicts how tall, in inches, a calf will be at 1 year old compared to calves of other sires. **Scrotal circumference (SC)** is measured in centimeters; bigger scrotums mean better bulls.

  You do not necessarily want a larger birth weight score because high birth weights often make it difficult for mothers to give birth. You most likely do want higher scores for weaning weight and yearling weight because high scores indicate the bull produces fast-growing calves. You have to weigh these scores based on your goals. If you want to sell calves at weaning, the most important score for you would be weaning weight. If you want to keep animals longer, the yearling weight score is more important.

- **Maternal:** This category predicts how a bull's daughters will perform if used for breeding. Traits measured in this area are:

  Calving ease maternal (CEM) predicts how easily a bull's daughters will give birth as first-calf heifers compared to other bulls' daughters. A high score means the daughters are more likely to deliver cows without difficulty.

  Maternal milk (Milk) predicts the mothering ability of a sire's daughters, specifically how her milk and mothering ability will translate to calf growth before weaning.

  Mature Weight (MW) predicts how much a bull's daughters will weigh (in pounds) at maturity compared to other bulls' daughters.

  Mature Height (MH) predicts how tall a bull's daughters will be (in inches) compared to the daughters of other bulls.

  Cow energy value ($EN). Not all cows require the same amount of feed to meet their energy requirements. $EN predicts how much a cow's energy requirements could save you in feed costs compared to the daughters of other bulls. $EN is expressed in dollar savings per cow, so a higher value in this category is preferable.

- **Carcass:** Carcass measurements are an indication of the yield from the final beef product. These consist of:

  Carcass weight (CW) predicts the hot carcass weight of a bull's offspring compared to other bulls. Hot carcass weight is a measure in pounds after an animal is slaughtered and its head, organs, intestinal tract, and hide are removed but before the carcass is chilled. Hot carcass weight is usually about 60 to 65 percent of the animal's live weight.

  Marbling (marb) is a fractional difference in USDA marbling scores. USDA assigns beef grades based on the level of marbling, or fat within the muscles. The marbling EPD compares the projected fractional difference in marbling scores from one bull's offspring to another's. It is expressed as a percent of one-third of a marbling score. *Marbling scores will be discussed in more detail in Chapter 9.* Higher marbling EPD scores are usually desirable.

Fat thickness (FAT) is measured at the 12th rib. Fat thickness is used to predict the overall fatness of an animal. Individual farmers will have to decide if they want this measurement to be higher or lower. Although fat is an indication of tenderness, this fat can also reduce the percentage of edible meat from the animal.

Ribeye area (REA) is a square-inches prediction of ribeye area for a bull's offspring. The ribeye area is where the most desirable steak cuts come from. The ideal ribeye area range is 12 to 14 inches, according research from the University of Georgia published in the *Journal of Animal Science*. Anything larger than this may be hard to sell.

- **Ultrasound:** Like the carcass scores, ultrasound scores predict the finished beef product. Ultrasound scores, however, are based on the results of an ultrasound image, so the data comes from the live animal rather than a carcass. Traits measured are:

  Ribeye area (RE) predicted in square inches

  Fat thickness (Fat) is a measurement in inches of fat between the 12th and 13th rib. It is calculated as 60 percent from the measurement at the 12th rib and 40 percent of fat at the rump. As in the similar fat thickness category in the carcass group, it is up to the individual farmer to decide if a high amount of fat thickness is desirable.

  Intramuscular fat (%IMF) predicts the difference in intramuscular fat in the ribeye muscle compared to calves of other bulls.

How to Handle Animals

Once you acquire animals for your herd, there will be many times when you have to walk among them and even move them from one place to another. This is especially true in a grass-fed operation that uses pasture rotation. Not only will you need to move your cattle from one section of pasture to another, or from the pasture into a milking facility, but you will also sometimes need to move your cattle from the pasture into your handling facilities, either for health maintenance tasks or to prepare them for sale. These animals are big, powerful, and quick, and many of them instinctively perceive humans as a threat. Although cattle herds become accustomed to farmers in their presence, learn the proper ways to move cattle and to behave around them.

Always be on your guard when you are among your cattle, especially when around bulls and mother cows. Farmers are injured or killed by bulls every year. If a bull charges you, run to a fence or some farm equipment to use as a barrier between you and the bull. Never turn your back on a bull. Mother cows can be aggressive when protecting their babies. Always keep the calf between yourself and the mom. If you get between them, the mother may try to knock you out of the way. If you have to work with a calf, separate it from its mother in a separate pen or behind another barrier, but make sure the mom can still see you and the baby. This will make the separation slightly less stressful for her. Return the calf to the mother as soon as you can.

Cattle want to be with their herd. If you must separate a cow from the herd and it does not want to go, try separating it with another cow to keep it company. Usually, it is easiest to just move the entire herd into your holding pen, and then separate out the sick animal and a companion. After you get the animals in quarantine, move the rest of the herd back into the pasture. The best way to move animals is to move slowly and deliberately and to use their natural instincts to your advantage. Every cow has a **flight zone**, a radius around it that it wants people to steer clear of. If you enter an animal's flight zone, it will move away. The size of the radius depends on the animal. Nervous animals have larger flight zones, which means they will not let you get as close as calmer animals before they move away. Calm or tame cattle have smaller flight zones and may not be as likely to move if you approach. Nervous animals will bob their heads or shake them back and forth, or they may paw at the ground while calm animals will continue chewing their cuds and will not move when you approach. If you move into a cow's flight zone, it will turn away so it can flee. If you do not want the animal to flee, take a step back out of the flight zone. Also learn an animal's point of balance, at about the shoulder. Use the point of balance to make an animal move forward or backward. Moving behind this spot will make cattle move forward, and moving in front of it will make them back up. Try to avoid a cow's blind spot, immediately behind them and about the width of their hips. You could startle an animal, and it could kick you or flee.

Talk to your animals and use specific commands because they will get to know your voice and will learn to recognize these instructions. Examples of cow calls are "Here, cow" or "Come, cow." Do not yell or strike the animals. Getting them excited will only make them more difficult to handle. A rattle paddle, a fiberglass pole topped with a large paddle with small balls inside, is useful for moving cattle. It is large enough that cattle can see it, and they will move away from the noise when you shake the rattle.

Use treats such as alfalfa pellets to encourage the animals to follow you at first. After a while, the promise of fresh grass will be enough to motivate the cows to follow you. They will recognize you as a source of fresh food and will be excited to see you. This makes other herd management tasks easier too because they will not be as reluctant to follow you into a new situation.

How to Introduce New Animals

Have a plan for bringing new animals, such as new bulls, replacement heifers, or stockers, onto your farm. A term for this type of plan is **biosecurity**, which means preventive steps you take to keep new illnesses from infecting your animals. When you purchase new animals, make sure they are transported to your farm in a clean vehicle and kept in a designated training pen. *Refer back to Chapter 3 about designating a training pen for new animals and a sick pen for sick animals.* Keep the new animals in the training pen, or quarantine pen, for at least two weeks — 30 days is optimal — before letting them in with the rest of your herd. If you notice some of your new animals are sick, isolate them further in a sick pen. It is permissible for the new animals to be visible to the animals that are not quarantined, but they should not be able to touch or to even get within a few feet of each other so potentially sick animals do not infect the animals that have been on your farm a while. These pens should have their own feeders or water troughs to prevent contamination of the food and water of the rest of your herd.

Move any animals showing signs of sickness into the sick area. If they show signs of sickness, take their temperature, and if any of them develop a fever or other illnesses, have your veterinarian check the animal. While these animals are quarantined, vaccinate them, treat them for parasites, or deworm them. *You will learn more about health management tasks in Chapter 6.*

Other strategies for preventing new animals from potentially infecting your herd include milking new animals last and sanitizing the equipment afterward, which you would do anyway. You can also wear different boots and clothing when working with animals in the quarantine pen. You can graze your animals near the sick corral, but do not introduce quarantined animals to the entire herd until a month passes to allow time to be sure none of the new animals are sick.

When to Cull

One way to improve your herd over time is to cull the underachievers so your breeding stock will consist only of the best performers. **Culling** simply means selling animals you do not want to raise anymore. These are animals you do not want to use for breeding or to continue to grow for beef or milk. When choosing animals to cull, look for signs the animal will not do well in your farm operation. If a cow shows signs of heat stress, such as rapid breathing, increased salivation, and panting, when your other animals do not, consider culling that animal. If it is not resistant to disease or parasites, it should not be included in your herd. A twin female is unlikely to breed; this is true if one of the twins is male or if both twins are female. Females not used for breeding can be finished for beef or sold at market to other beef producers.

Animals you do not want to breed or finish for beef can be sold at sales barns to farmers who will continue to raise them for beef. Also, sell sick or old animals at these auctions. These are the ones that usually become fast-food hamburger or other lower-quality products. If a problem is obvious, such as cancer eye, the auctioneer will announce the problem during the auction. Here are some criteria to consider when culling:

- **Temperament**: Temperament is an important factor to consider. A high-strung or nervous animal will be high maintenance and, in the end, their meat probably will not be good quality. Frequently stressed animals produce hormones, including adrenaline, that make their meat tough. It is normal for animals to be wary of you at first, but it is not normal for them to try to jump the fence every time they see you. If they are this nervous, you must get rid of them. You do not want aggressive animals, but you also do not want shy animals because they often get crowded out of feed or water and will not get all the nutrition or water they need, which will keep them from growing as fast or producing as much milk as they should.

- **Fertility**: If your cow shows signs of infertility or other reproductive issues, she should be culled. If you put a cow with a healthy bull for 60 days and she is not pregnant, there is something wrong with her.

Those that breed late can also culled. Occasionally, if a heifer fails to breed, known as an **open heifer**, some farmers will keep her around for another season because buying a replacement would be expensive. Usually, because it is expensive to feed a heifer that may have trouble breeding the following year, open heifers are culled.

- **Prone to disease**: If an animal gets sick and you cannot get it healthy, sell it at an auction. It will likely be bought by a large-scale beef producer and will probably end up as fast-food hamburger. If you cannot get rid of worms in two treatments, get rid of the affected cow. If it has chronic eye problems or shows early signs of cancer eye, it should be culled. *Cancer eye will be discussed in more detail in Chapter 6.*
- **Performance**: Calves with low weaning weights should be candidates for early sale. If a cow's first couple calves have low weaning weights, it might be best to sell her at auction too.

It takes time to build a good herd. Good farmers continuously monitor their animals to identify their best producers and the ones not suited for success in a grass-based system. As you gain experience buying, selling, and breeding animals, the quality of your herd will continue to improve. *In the next chapter, you will learn how to take care of these animals.*

• Chapter 6 •

Animal Health and Nutrition

TERMS TO KNOW

Milk letdown: The process of releasing milk into the udder

Healthy animals will produce the best results on your farm. Cattle in good health will grow more quickly and produce better meat than sick or stressed animals. Healthy animals will also produce larger quantities of better-quality milk. Take steps to help your animals grow and stay healthy with a herd health plan that includes vaccinations, routine checkups, and treatments when symptoms of illnesses appear. Also, know how the digestive systems and udders of your cattle work.

Digestive System

A cow's digestive system is the engine that makes your whole farm function. This is where the cow uses the nutrients from your grass to grow or to produce milk. Both processes are key to a profitable cattle farm.

Cows use their tongues to grab grass and then tear it off with their front teeth, called incisors. A cow only has lower incisors and has a hard pad in place of upper incisors. They chew with their back teeth, called molars, and swallow this wad of food, which travels down the esophagus to the stomach. Cows have four stomachs that add up to a 55-gallon digesting machine. The four stomachs are, in order from first to last, the rumen, the reticulum, the omasum, and the abomasum. The first stomach, the rumen, contains organisms that break down fiber in grass and turn it into volatile fatty acids and other useful nutrients. The second stomach, the reticulum, helps the cow regurgitate wads of food back into its mouth for more chewing, which helps the microbes in the rumen by further breaking down the tough grass. The reticulum also acts as a storage area for any indigestible foreign objects the cow swallows, such as screws or anything else lying around in the field. Food that has been broken down well travels on to the omasum, which absorbs more useful components from the food before sending it on to the fourth stomach, the abomasum. The abomasum is the stomach that works most like a human stomach. After the abomasum digests and absorbs nutrients, the food moves on to the intestinal tract.

A newborn calf does not have a functioning rumen, and it takes about four months for the rumen to develop. When a calf drinks milk, folds of tissue make a groove from the esophagus to the abomasums, bypassing the rumen. Introducing some roughage at an early age helps the calf develop the rumen. If you leave them out in the pasture, they will try grazing on their own. This immaturity makes calves susceptible to dangerous bout of bacteria-caused diarrhea know as scours. *You will learn more about this later in this chapter.*

As mentioned in Chapter 1, the chemistry in a cow's rumen is hospitable for bacteria that aid in the breakdown of grasses. The starches in a grain-based diet change the chemistry in the rumen and make it more acidic. This excess acidity means grass can no longer break down, and the animal will get no nutrition from the grass. Also, this excess acid is hard on the animal's liver and can eventually cause liver failure.

Bloat

You also have to be careful about feeding cattle pastures predominantly made of legumes or especially lush spring pastures. Although these are healthy for the animal, if a cow eats too much too quickly, it will bloat. Bloat is caused when froth builds up from rapid fermentation caused by the microbes in the rumen. This froth traps gases that normally escape through belching. Never turn a hungry cow over to a fresh legume-based pasture. Instead, introduce the cattle to a small amount of legume-based pasture for an hour or so at a time over the course over a few days, preferably after it eats grasses for much of the day. Feed them hay or other grasses first so they will be less inclined to gorge themselves.

You can also purchase bloat inhibitors from veterinary offices or feed mills that you administer in water, through pills, or in licks, edible blocks containing ingredients that suppress the formation of bloat. Cattle are more at risk for bloat if they eat from lush new pastures after being used to dry or poor-quality pasture. To prevent bloat, do not turn them loose on pastures still wet from rain or dew. The risk of bloat from alfalfa is even greater after frost, so wait a week after a frost to graze alfalfa. The risk of bloat from alfalfa decreases as the plant matures past the vegetative state. A cow with bloat may show a swelling on its left side where the rumen is located, and it may also make distressed noises, kick at its stomach, or have trouble breathing. If a case of bloat is mild, you may be able to help the animal walk it off by walking alongside it and leading it around the pasture. More serious cases may require a stomach tube and anti-frothing agent. If you do not know how to do this, a veterinarian can help you. Severe cases can kill animals quickly.

Udder Anatomy

Dairy farmers should understand how the udder works because the udder is the factory for all dairy products because it produces milk, butterfat, and protein. The udder is four teats and a sack divided into two front and two rear quarters. From behind, you can see it is divided in half by an indentation called the mammary groove. The teats, at the ends of each quarter, are release valves for milk for suckling calves. They are smooth and hairless. The duct through which milk exits is called the streak canal, which relaxes during milking to allow milk release. Using a germicidal postmilking solution on the teats will help keep bacteria from invading in the several minutes before the duct closes. The solution will also keep flies away. A strong suspensory system of ligaments and tendons provide support and keep the udder in proper alignment with the body. If this supporting structure breaks down, the cow is more prone to teat or udder injury and more susceptible to the udder infection called mastitis. *Mastitis will be discussed in more detail later in the chapter.*

The udder has a vast vascular system. Large mammary arteries are located on the side of the udders. Ducts and cisterns, storage areas for fluids inside the udder, drain the milk from the secretory tissues, which make up the majority of the udder.

The process of releasing milk into the udder is called **milk letdown**. It is brought about through the release of a hormone called oxytocin at the base of the cow's brain. Oxytocin affects the muscle cells surrounding the udder's milk, which causes them to contract. This squeezes the milk out of the cells and into the milk ducts that drain to the teat. Stimulate the release of oxytocin by gently massaging the udder for 15 seconds. This will cause the teats to fill with milk within a couple of minutes. Cattle need to be calm during milking or they may not release oxytocin. Stressed cattle release adrenaline, which blocks oxytocin for up to 30 minutes. This decreases milk yield and leads to milk retention, another cause of mastitis.

Necessary Nutrition

A cow needs about 26 pounds of dry matter intake (DMI) each day. Cattle also need this daily DMI to obtain the necessary amounts of various nutrients. The biggest of these necessities is protein. Cattle usually require about 7 to 14 percent crude protein in their DMI, according to the ATTRA publication "Cattle Production: Consideration for Pasture-Based Beef and Dairy Producers." Dairy cattle, cows in late-term pregnancy, lactating beef cattle, and pregnant heifers have higher nutritional needs than animals such as stocker cattle. Cattle that are not growing, such as early-term pregnant cows that reach maturity or healthy bulls, have lower nutritional requirements.

Other nutrients such as calcium and phosphorous are also important for cattle health, though cattle only need less than 1 percent for each nutrient. Have cut pasture grasses, hay, or silage analyzed to learn the nutritional content of your herd's food supply. The National Forage Testing Association certifies labs to ensure the accuracy of forage tests. See a list of these labs on its website (**www.foragetesting.org**). If you get your forage analyzed, the key measurements are total digestible nutrients (TDN) and crude protein (CP). Forage with 10 to 13 percent CP and 55 to 60 percent TDN should provide adequate nutrition for all members of your cattle herd, according to ATTRA. Typical pastures in the vegetative stage, and even the boot stage, meet these benchmarks. Legumes in the vegetative state can contain more than 20 percent crude protein.

On high-quality pastures, cattle may need to be supplemented with fiber such as soybean hulls to help them more efficiently use protein. If pasture is lower quality, cattle may need to be supplemented with protein such as cottonseed meal. Feed these supplements by either leaving them in the pasture or by using portable feeders such as troughs you can move around the pasture.

Cattle can get most of their nutritional needs straight from pasture grasses. But you will also have to provide your cattle with mineral supplements to replace salt lost in urine, help produce milk and develop fetuses, and help with general body functions. These are generally given in a pre-mixed ration, either as loose granules or in block form, placed in mineral feeders. You may need different mixes during different seasons. For example, in spring, when grass grows quickly, cattle often do not get all the magnesium they need, so they will need a higher percentage of magnesium in their mix. Selenium is also commonly deficient in soils. Local universities or feed stores can recommend a mineral mix for you.

In winter, many farmers use energy licks to supplement hay and dormant pastures. These licks, which can be put in the pasture so animals are free to eat from them, consist of a protein source, an energy source, and usually molasses. They can be formulated into 40-pound blocks up to 200-plus-pound tubs, and they are left in the pasture for cattle to lick. Purchase these licks at farm feed stores. You have to be careful and read about the ingredients in these licks because sometimes they use animal byproducts, which are often avoided by grass-fed farmers because they are not a natural diet for cattle. Feeds containing animal byproducts are prohibited by the American Grassfed Association. You can find energy supplements that will be acceptable for your program. *Lists of supplements approved and banned by the AGA can be found in Appendix B.* The most up-to-date AGA standards can be seen at **www.americangrassfed.org**.

Body Condition Score

Beef farmers can gauge the health of their animals by using a body condition score system. This score is used as a guideline to help you identify healthy, overweight, and undernourished animals. Monitoring the body condition score will allow you to spot animals in danger of slipping to unhealthy levels, which reduces fertility. If a heifer is in danger of losing her body condition, supplement her feed or even wean her calf early. You may also wish to cull cows that have difficulty maintaining a desirable score. Some farmers have the whole system memorized; others simply learn to recognize a fat or undernourished animal and adjust its feed accordingly.

There are separate scoring systems for beef and dairy cattle. Beef cattle are measured on a system from 1 to 9. A score of 1 means the animal is nearly starving; a score of 9 means the animal is obese, and scores between 5 and 7 are optimal. Ohio State University has pictures of animals in each condition at **http://ohioline.osu.edu/l292/index.html** in its fact sheet "Scoring Animals Can Improve Profits."

Here is an overview of each number in a typical beef body condition scoring system:

1 is emaciated, with no fat left and very little muscle left.	The animal may barely be able to stand. An animal in this condition is probably suffering from a disease or parasites.
2 is very thin.	You can see the bones of the animal, and there is quite a bit of muscle loss.
3 is thin.	You can see the animal's foreribs, roughly in the center of the animal. You can feel each point of the spine.
4 is still thin but closer to being a healthier weight.	There is some fat on the animal and the spine is not obvious to the touch.
5 is not fat or thin.	The animal is starting to flesh out around the ribs, shoulder, and tail.
6 is good.	There is some fat on the back, the hips, and the brisket, the front area below the neck.
7 is the best condition.	The animal has fat on its back and the base of its tail. This fat can be used as a stored reserve in winter.
8 is fat.	The bone structure is no longer visible.
9 is extremely obese.	The animal has a fat udder and patchy fat around the tail head.

Heifers at calving should be at about 6. Heifers with more fat than this are more likely to have reproductive difficulties or problems calving. Make sure they maintain a score around 5 or 6 after calving until it is time to breed them again. Mature cows are not growing and can calve successfully with a score of 5.

You do not want a fat bull because this lowers fertility, especially if fat collects in the scrotum. Keep him within a range of 5 to 7. Bulls have high nutrition requirements during the breeding system. You have to watch them to make sure they do not lose weight. Bulls expend more energy than they take in while they concentrate on breeding. Perhaps, give them nutritional or energy supplements during this time period.

There is a similar scoring system for dairy cattle, with 1 being thin and 5 being obese. Dairy cattle should usually be between 2.5 and 4. Judging a dairy cow's score under this system depends more on the appearance of the flesh around the pin and hook bones and tail head, visible on the back of the cow. Hook bones are at the corner of the top of the rump, and pin bones are at the bottom of the pelvic area near the midway point of the tail. The tail head is the base of the tail. The University of Arkansas published a guide, "Body Condition Scoring With Dairy Cattle" that has pictures of cows in each condition score (view it at this address: **www.uaex.edu/other_Areas/Publications/PDF/FSA-4008.pdf**). Here is an overview of each point in a typical dairy body condition scoring system:

1 is a very undernourished animal.	The pin and hook bones are visible, and there is a cavity around the tail head. You can see individual ribs.
2 is also undernourished.	The ribs, pin and hook bones, and cavity around the tail head are not as visible, but you can easily feel these bones if you touch the animal.
3 is good condition.	The area around the pin and hook bones is smooth.
4 is a fat animal.	All bones are covered with fat, and the area around the pin and hook bones is flat. Cows or heifers that score above 4 could have calving difficulties.

5 is obese.	There are fat folds covering all bones and the tail head.

Herd Health Tasks

There are many health-related tasks you will have to perform routinely, such as vaccinations and pregnancy checks. These tasks require you to move your herd into the holding area, which can be stressful for the animals. For this reason, schedule as many herd maintenance tasks together as possible. For example, you could vaccinate in the fall when you also check cows for pregnancy. You may also need to enlist the help of a friend or veterinarian for these tasks. For most of these tasks, you will have to restrain the animals in your head gate, and many head gate systems require an operator who will close the gate around the animal's neck after another person leads the animal to the gate.

Tagging

A common practice is to use ear tags to help identify the individual animals in your herd. These tags hang from the ear and feature an easy-to-see number. Tagging animals only takes a few minutes. You simply use a tool called an ear tag applicator to poke a hole in an ear and attach the tag, sort of like an earring. Tag them when they arrive at your farm, shortly after they are born, or when you perform other tasks such as vaccinations or castrations. Tags help keep records of your animals, including which calf was born to which mother, and also aid in identifying animals for your vaccination and other medical records. If you cannot tag calves early on, you will still be able to keep track of which calf belongs to which cow. Calves stay close to their mothers, and if they get separated, the mothers do a good job finding them. If the mother will not let you tag her newborn calf, separate them after six to eight weeks and put the calf in a head gate for tagging.

Castration

Most male calves are castrated. Only the best males are used as studs, and those not lucky enough to be picked are worth more as steers, used for meat rather than breeding. Uncastrated males are a threat; they might attack farmers, they will be more likely to fight each other, and they might impregnate nearby cows. Calves can be castrated soon after birth, but many farmers wait until the calves grow to a few hundred pounds because the extra testosterone increases early growth. Waiting two to four months before castrating your bull means an additional 50 pounds or so at weaning. You can castrate older calves, but it takes them longer to recover because the arteries and nerves to the testicles are larger than in young calves.

The calf should be restrained in a head gate. Because the calf could kick during the procedure, castration is a two-person job. One person lifts the tail, not hard enough to lift its feet off the ground, but hard enough to keep the tail straight up, to prevent it from kicking. If you have never castrated a calf before, it is a good idea to enlist the help of a veterinarian or an experienced farmer.

For calves up to 2 months old, use band castration. In band castration, a tool called an elastrator is used to stretch a sturdy rubber band around the scrotum. This is initially uncomfortable, but the tissue becomes numbed by the tight band, which cuts off the blood supply to the scrotum. Within two weeks, the scrotum and testicles fall off, leaving a fully healed scar. If they fall off in the field, you will probably never see them again.

For calves older than 2 months old, use a tool called an emasulatome, also sometimes called a clamp, an instrument with blunt crushing jaws to damage the testicular cord without hurting the scrotum. To get to the testicular cord, lift the calf's tail, grasp the testicles between its legs, and pull them taut. You then apply the emasulatome to each testicle's spermatic cord one at a time to crush them.

There is also a surgical procedure for castration used on older calves. In this procedure, a scalpel is used to cut off the bottom third of the scrotum. Then, one person grabs a testicle and pulls it to break the attachment that secures it to the inside of the scrotum and crushes the spermatic cord with the emasulatome. The spermatic cord is then cut below the crushed area. Repeat this with the other testicle. The testicles are usually discarded, but they could also be kept and cooked as food — you may have heard the terms "calf fries" and "Rocky Mountain oysters," among others.

Vaccinate at the same time as this procedure to prevent infections from tetanus and the blackleg-causing clostridium bacteria. *This will be discussed later in this chapter.* As you might guess, this is a painful procedure for the calf. It will lose much of its appetite for a couple of weeks. Observe the calf closely for a few hours to make sure there is not excessive bleeding. If there appears to be a problem, call your vet. Fly season is June, July, and August, so time castration after this so flies do not bother the wound.

Dehorning

Dehorning is another routine health-care task. Horns are often undesirable because they are dangerous to humans and can be harmful to other cattle. For this reason, horned cattle do not command full price at sales barns. Horns grow from horn buds, two small bumps located on each side of the forehead. Watch someone else dehorn animals before you try to do this task on your own. Dehorning is painful for the animals, but your veterinarian can help you apply local anesthetics or painkillers to reduce the discomfort. The younger the calves are when you dehorn them, the quicker their recovery time. The best time to dehorn is soon after birth. You have a few options for dehorning:

- An electric dehorner is a quick and easy method. Firmly apply the heated end of the dehorner to the horn bud to kill the horn tissue. A ring of char will form on the horn growth area, and it will scab over. The horn tissue will fall off after a few weeks.

- Scoop dehorning involves using a dehorning spoon or tube to gouge out the horn buds. All methods of dehorning are painful, but this method seems to cause more pain than the others.
- Use caustic paste from farm supply stores for a paste application. First, isolate the calf. Then, trim off the hair around the buds and apply a thin layer of petroleum jelly to the skin around the horn to prevent it from burning. Wear gloves and be sure not to get the paste in the calf's eyes. Use a wooden applicator to put the paste on the horn buds to kill the horn tissue. The paste application stings, and the calf may try to rub it off. Cover the horn buds with patches or duct tape to keep the calf from injuring itself or others with this rubbing, according to Neil Anderson, a lead veterinarian with Ontario Ministry of Agriculture, who writes for TheCattleSite, a free online information resource (**http://theCattleSite.com**). The duct tape should fall off on its own in a few days.

Use any of these methods until the calf is 2 months old. If the calf is older than that, you will need to use a Barnes dehorner. A Barnes dehorner has sharp, hinged blades to slice off the horn at its base. There will probably be lots of bleeding, and the arteries will need to be pulled to stop the bleeding, which is why a veterinarian should dehorn older calves to ensure bleeding is controlled.

Vaccinations

A vaccination program helps protect against viruses, which are difficult to treat. It is easier to sell vaccinated cattle because buyers will be confident your animals will stay healthy. Vaccines come in two versions. The first type of vaccine you can give your cow is a modified live vaccine, which means the disease-causing organism is still alive but modified to not cause the disease in healthy animals. Live vaccines are better at stimulating the immune system, but they are unsafe for pregnant cattle or to calves still nursing. The alternative is a killed vaccine, which means the disease-causing organism is dead but can still stimulate the immune system.

Before calves are weaned, they should be vaccinated with a seven-way blackleg vaccine, a five-way leptospirosis vaccine, and with a combination vaccine for the diseases IBR-BVD-PI3. Heifers should be vaccinated against the disease brucellosis between 4 and 12 months old. *You will learn more about these and other diseases in more detail later in this chapter.* If your veterinarian says pinkeye is a problem in your area, all your cattle should be vaccinated against this prior to the start of the grazing season. Many vaccinations, such as those common for pinkeye, require booster shoots within a few weeks after the first vaccination. Consult with a vet to develop a vaccination plan. Vaccines are permitted under organic and AGA standards.

Sick Animals

Pasture-based feeding systems reduce many of the problems associated with factory farms. The natural diet eliminates the unhealthy levels of stomach acid caused by grain feeding, and the animals live longer because they are not pushed to unnatural levels of growth or milk production. Animals that live in pastures also experience fewer hoof problems and respiratory illnesses associated with feedlot or dairy barn confinement.

Signs of sickness

A healthy cow has clear eyes, erect ears, a healthy coat, clean muzzle and nostrils, and a good appetite. A sick cow will look or act differently than the rest of the cows and will often have droopy ears, arched back, or a limp. It may also pull away from the herd, act listless, spend most of its time lying down, or lose its appetite. It could have a higher-than-normal temperature. A normal temperature is 101.5 degrees, and a sick cow's temperature will be more than 103 degrees.

Handling a sick animal

You must isolate sick animals from the herd so they do not spread the sickness to other animals. As discussed in Chapter 3, you may want to have a sick pen to quarantine sick animals. The best thing you can do for cattle with viruses is make them comfortable. Make sure they have plenty of food and water. You may also be able to use medications to reduce fevers and inflammations. Most sick animals will get better; if they do not, sell them at a stockyard auction to a mass beef producer.

Types of Treatments

Form a relationship with a veterinarian you trust. You can find veterinarians through referrals or even by calling a few to see if you find one you prefer. Find one who will make emergency calls in the middle of the night. Your vet should understand the goals of your health program, especially if you want to adhere to organic principles. Conventional treatments involve synthetic pesticides and parasiticides and include antibiotics for many illnesses; these are not allowed in organic programs.

Organic production also prohibits pharmaceutical medicines and synthetic insecticides. Antibiotics are not permitted under organic or AGA standards except when the animal's life is in danger. However, the treated animals must be removed from the certification programs. Animals given antibiotics cannot be used for beef or dairy production under these certification programs. On the National Organic Program website (**www.ams.usda.gov/AMSv1.0/nop**), refer to section 205.603 of the NOP regulations for a list of synthetic substances allowed for use in organic livestock production. If you have questions about any substances, check with your certifying agent or your veterinarian.

As you gain experience, you will have fewer reasons to call a vet because your herd management practices will encourage healthy animals. You will also become more confident in your own ability to recognize and treat illnesses.

Organic and homeopathic treatments

Producers who strive to raise their animals as naturally as possible often eschew pharmaceutical treatments for organic or homeopathic treatments. **Homeopathy** is a system of treatment that uses natural substances including garlic, herbs, vitamins, and oils. These substances are given in tiny amounts to trigger the immune system to fight illnesses. If these substances were given in large quantities, they would produce symptoms similar to the sicknesses being treated. Find these products at specialty stores, which often sell products online. Two such companies are Crystal Creek (**www.crystalcreeknatural.com**) and Agri-Dynamics (**www.agri-dynamics.com**).

Some veterinarians use acupuncture on cattle to treat problems including fertility issues and digestive issues. The International Veterinary Acupuncture Society, a nonprofit group that promotes veterinary acupuncture, has a website (**www.ivas.org**) that lets you search for veterinarians who perform acupuncture by state. Homeopathic treatments and acupuncture for cattle are divisive practices. Many people deride them as quackery, while others swear by them.

Although organic producers strive to stay away from conventional treatments, cattle sometimes contract severe illnesses that do not respond to alternative treatments. In these situations, you must treat them conventionally to save their lives. After you do this, you must be sure not to sell them under labels such as organic that prohibit these substances.

Common Illnesses

There are many diseases common to cattle, most often caused by bacteria or viruses. Stressed cattle, such as calves being weaned or sold at auction, are most susceptible. Reduce the number of illnesses on your farm through a herd health program that includes vaccinations, good nutrition, and parasite control. A good pasture rotation will also keep the animals ahead of many parasites and illnesses they pick up from animal waste left in the pasture. This section includes information about various cattle diseases. Your veterinarian will advise you on sicknesses to vaccinate for. Common vaccinations are for IBR, PI-3, BVD, brucellosis, and blackleg.

IBR: Infectious bovine rhinotracheitis (red nose) is a respiratory illness that causes nasal discharge, mouth ulcers, a red muzzle, and a temperature of 106 degrees or higher. IBR spreads rapidly through unvaccinated herds.

PI-3: Parainfluenza 3 causes typical flu symptoms in cattle, including coughing, fever, and eye and nasal discharge. The cow will probably also lose its appetite. Most cases are mild, but these infections can eventually lead to pneumonia.

Pneumonia: Pneumonia is an inflammation of the lungs. It is most common in calves and can start with viruses that lead to secondary infections. Symptoms of pneumonia in cattle are rapid breathing, coughing, watery eyes, nasal discharge, and lack of appetite. Antibiotics are the most common treatment for pneumonia. You can also buy various tinctures, alcohol-based liquid extracts of natural herbs, to treat pneumonia organically and homeopathically.

BVD: Bovine viral diarrhea (BVD) is diarrhea including symptoms such as dehydration, fever, mouth sores, and rapid breathing. Cattle of all ages can become infected with BVD, but those 24 months or younger are most susceptible.

Brucellosis: Symptoms of brucellosis include arthritic joints, aborted fetuses, and afterbirth not expelled after calving. "Abort" is the industry terminology when a cow miscarries her unborn calf. Bulls infected by brucellosis are usually sterile. Brucellosis is contagious to humans and causes people to experience fevers, fatigue, joint pain, headaches, and psychotic behavior. Brucellosis is all but eradicated from cattle in the United States due to vaccinations and milk pasteurization, but it does surface occasionally in wild bison and elk. Brucellosis is caused by the bacteria *Brucella abortus*. Raw milk supporters argue the threat of contraction to humans is extremely low because there have been so few cases in cattle in recent decades. About 100 people in the United States contract brucellosis each year, according to the Centers for Disease Control and Prevention (CDC), but most of them catch it while traveling to other countries.

Leptospirosis: This is a disease that affects adult cows' reproductive organs, causing infertility, stillbirths, and abortions. Calves may also become ill with fever and blood problems or suffer reduced milk production. There are many strains of leptospirosis, but five types of bacteria cause most illnesses. The bacteria are spread through urine and manure and live in the soil for a few weeks after being deposited. Vaccinations build resistance to the strains that cause the illness. In healthy herds, infected animals will show no external signs. The disease can be spread to people and pets, according to the CDC, usually manifesting as flu-like symptoms.

Blackleg: Blackleg is a disease that most often affects young calves and can kill an infected animal without warning. There could be no signs of infection, but one day, you may just find the calf dead. The disease is caused by spore-forming *Clostridium chauvoei* bacteria, which lie dormant in the soil for years and infect a calf that eats contaminated plants. The spores lodge in various organs and muscles in the calf, and they multiply when an injury occurs. It does not even have to be a severe injury — even slight bruising can trigger the multiplication.

Winter dysentery: Winter dysentery causes watery, explosive, dark diarrhea, intestinal obstructions known as colic, cough, decreased milk production, and lack of appetite. The disease usually clears up in a couple of weeks, but be sure sick animals have plenty of water, feed, and mineral supplements to replace losses from diarrhea. The cause of winter dysentery is not certain.

Johne's disease: Johne's disease is an infection of the intestinal tract that is fatal if it progresses to its final stage. It is caused by the bacteria *Mycobacterium paratuberculosis*, which prevents animals from absorbing protein properly, eventually leading to diarrhea. Johne's disease often appears similar to other diarrheal or weight-loss diseases, including muscle wasting, weight loss, dehydration, and loss of appetite. If you suspect Johne's disease in any of your cattle, take fecal cultures or environmental samples from your field to determine if *M. paratuberculosis* is present. The bacteria are spread through fecal-oral contact, such as infected grasses or pond water, or it can be transmitted intrauterine and through the milk. It can take two years to reach its final stage; once it does, the disease will result in death. Many animals in the herd may have the disease but not reach the final stage. There is no treatment for Johne's disease.

Scours: Scours is the name for diarrhea in very young calves. It can be caused by many of the infections listed here, including BVD. A calf with scours may appear weak or lethargic. Scours causes the calf to be dehydrated and it will endanger the calf's life. You must isolate the calf from the herd and feed it electrolytes. If the calf is healthy enough, bottle-feed it the electrolytes, which you can purchase from your vet or from feed stores. Be sure you get a type intended for calves as opposed to other animals. If it is too sick or refuses to eat, you will have to force-feed it electrolytes with an esophageal tube, a plastic tube attached to a bag holding the electrolyte formula. If you have not treated scours before, it may be best to call a vet and ask for help because you may only have a few hours before it is too late.

Pinkeye: Pinkeye can be rapidly contagious. It affects calves more often than adults. Animals with light-colored faces, such as Herefords and their crosses, seem to get pinkeye more often than other breeds. Pinkeye, also known as infectious bovine keratoconjunctivitis (IBK), is caused by the bacterium *Moraxella bovis*. Contributing factors include exposure to sunlight, dust, flies that bother the face, and an infection of the IBR respiratory virus. Traditional treatments include pesticide-laced ear tags and antibiotics. There are also organic or homeopathic treatments, including pellets and eye washes. You can also use an eye patch to protect the infected eye.

Cancer eye: Cancer eye is another illness most common in white-faced cattle such as Herefords. Cancer eye, or squamous cell carcinoma, is the most common cancer in cattle. Sunlight, dust, and aging are all factors leading to the development of cancer eye. A smooth, white lesion around or on the eye often appears in early stages. In advanced stages, there may be a growth of rough, bumpy tissue that can quickly erode and die. Your veterinarian can attempt removal at an early stage, but the cancer often recurs.

Rabies: Rabies is a threat to all warm-blooded mammals, including cattle. Cattle can contract the disease when bitten by infected animals. They can also transmit this disease to people through their saliva. In cattle, rabies presents itself in two ways: a furious form and a dumb form. Cattle with the furious form may charge people, objects, or other animals; bellow frequently; and run around frantically. The dumb form causes infected cows to act depressed, drool, and become paralyzed. If you suspect rabies, call your vet to come check it out. It is usually possible to quarantine an infected cow before it infects other animals. Rabies is fatal and has no treatment available. Most cases of rabies involve wild animals; the disease is rare in domesticated animals. A vaccination is effective in preventing the disease, though it is not given in all areas because of its cost. Sometimes veterinarians will recommend vaccinations at times when many wild animals in an area have tested positive for rabies. A person bitten by a rabid animal should thoroughly wash the wound for several minutes and seek medical attention immediately after that. You cannot get rabies by drinking pasteurized milk or eating cooked meat from a rabid cow. It is theoretically possible to get rabies by drinking unpasteurized milk from a rabid cow, but there are no documented cases of this.

Grass tetany: Grass tetany is a disorder that affects the nervous system. It is a deficiency of magnesium in the blood that most often affects lactating beef cattle grazing on young, greening pastures low in magnesium. Symptoms include staggering, muscle tremors, convulsions, hyper excitability, eyelids that snap open and closed, and convulsions. Grass tetany is often fatal and can spread throughout the herd, so it must be treated early. Move infected cattle away from the herd, keep them quiet, and call your veterinarian, who can give them magnesium in an injection or through an IV. Follow-up treatment may include oral or rectal supplements or injections just under the skin.

Milk fever: Milk fever, or hypocalcemia, is a depletion of calcium in a cow's blood stream because of the huge demands for calcium in milk production. It can occur soon after calving, especially in older dairy cattle. Calcium is necessary for muscle contractions, so a cow with milk fever is weak and wobbly, will have cold ears, might be restless, and may pace or bellow. Treat milk fever with calcium given in the vein and possibly a supplemental oral calcium gel. Buy calcium at farm supply stores or from your veterinarian.

Mastitis: Mastitis is the biggest health problem of the dairy cow because it decreases milk production and can make the milk undrinkable or unsalable. Milk from the affected cows may be watery, chunky, or have flakes. Mastitis is usually caused by a bacterial infection. Signs of mastitis include swelling, pain in the udder, and heat — the udder will feel hot to the touch. Severely sick cattle will lack an appetite, be weak, and may not be able to stand. The chances of such infections are increased with poor milking techniques, faulty milking machines, or teat injuries and sores. Many cases are caused by *Staphylococcus aureus*, a hard bacterium to eradicate. Mastitis can also spread from cow to cow in dirty living environments through streptococcus bacteria. A release of toxins from coliform mastitis can cause a cow to become extremely ill and so weak it may not be able to stand. It will stop eating and cease to have rumen contractions. In this case, call a veterinarian to promptly treat this type of mastitis in order to save the cow's life.

Other cases may not be as severe and therefore not obvious, but they will be chronic. Signs of these cases include increased somatic cell counts in the bulk tank. Somatic cells are white blood cells that increase when the immune system fights bacteria. The somatic cell count increases in relation to the severity of the infection. Test for these cases with a California Mastitis Test (CMT) kit, which you can buy at farm supply stores, through online farm supply companies, or at livestock veterinarian offices. This kit comes with a paddle with four shallow wells. Milk from each quarter is stripped into its corresponding well, and a solution is squirted into the milk and swirled with the paddle. The solution will react with the mastitis and cause gelling.

Reduce the chances of mastitis with good milking procedures, good milking machines, and a clean environment. Prior to milking, clean the udder with a soft, disposable dry cloth to remove any dirt and debris. If the teats are dirty, they should be cleaned and lightly disinfected with an iodine cleaner and then dried well before milking.

After milking, each teat should be dipped in a post-milking dip solution. This solution is usually an iodine-based liquid and can be readily found in any dairy supply store. There are also organic teat dips available. Cows in grass-based dairies are allowed back in the pasture after milking. Controlling flies and other insects is another way to cut down on mastitis, as flies can carry some of the bacteria that can cause mastitis.

Hoof health

Large commercial dairies create many hoof problems because the animals are forced to stand on hard, often dirty concrete. Although cattle raised on pasture should be able to get more exercise and have healthier hooves, they still sometimes have foot problems. Cattle with painful hooves will not eat well and can develop other leg problems. A normal cow's hoof is rounded at the tip, and the skin between the toes is pinkish-white. An unhealthy hoof is overgrown at the tip and sides, and the skin between the toes may ooze fluid from infection, have reddened tissues, or may have a foul odor. Help heal hoof problems by trimming the hoof. Do this yourself, or hire a professional hoof trimmer or veterinarian. A professional will check for injuries, infections, ulcers, or warts and will let you know the best way to treat the injury.

Pests and Parasites

There are many pests that affect cattle either by irritating them or by acting as parasites, which siphon away the animals' nutritional intake. Flies, worms, fleas, and lice are the most common pests your herd will encounter. Minimize the impact pests will have on your herd with a good health program. Internal pests can spread from one animal to another when their eggs are deposited into the pasture through manure. The insects or larvae can make their way up grass blades where they will be eaten by a new host. A good pasture rotation fights against flies because cattle are moved away from fly eggs before they hatch. Another natural way to combat flies and other pests is to rotate chickens in behind the cattle because the chickens will pick out the fly larvae.

Flies: Flies are extremely irritating to cattle and carry diseases they introduce from contacting cows' eyes or wounds. Flies are especially a problem in areas where manure concentrates and animals linger. One type of fly, the warble fly, will lay eggs on cattle, and the larvae will burrow and settle underneath the surface of the skin. There are a few ways to control flies: Hang burlap bags of the natural compound diatomaceous earth, a soft rock that can be turned into a fine powder. You can buy bags of this powder in many places, including common department stores. Cattle will rub against these and coat themselves with the fly-repelling substance. Diatomaceous earth is also reported to be effective on fleas and lice. To deal with internal parasites, some farmers feed Diatomaceous earth to cattle by mixing it into feed supplements. There are also fly traps available that can help eliminate the insects.

Worms: Worms are parasites that often cause malnutrition. There are many species of worms that affect cattle. The more common types include the brown stomach worm, lungworm, and liver fluke. Infected animals pass on the eggs through manure, and these eggs can hatch in the field. Larvae then live on grass or in soil and can then infect other grazing animals. Sellers of organic and homeopathic remedies will also sell natural dewormers. Many farmers deworm regularly; the number of times per year varies by farmer. Your veterinarian can help you devise an effective strategy for worm prevention.

Cattle lice: Cattle lice are a common problem. Lice bite, suck blood, and lay eggs on the cattle. They cause itching and discomfort and put the animals at risk for disease because they are stressed. Conventional treatments include topical insecticides. Spraying cattle with neem oil, a vegetable oil used as a natural insecticide, is one natural way to prevent fleas and lice.

Fleas: Fleas bother cattle in the familiar ways they bother family pets: causing itching and discomfort. As with lice, insectides and natural treatments such as neem oil are ways to treat for fleas. Products made from neem oil are available at farm supply stores and organic specialty stores.

Euthanasia

There may be times when you can do nothing else for an animal besides relieve its suffering. Serious injuries, certain calving complications, or incurable illnesses can happen in any herd. If an animal cannot get up, it may have to be euthanized. This could happen after a calving injury or with fractures to weight-bearing areas such as legs or hips. If an animal cannot stand because of illness or injury, call a veterinarian who can determine if it can be saved or if it would be more humane to euthanize the animal. Experienced farmers learn to recognize when there is nothing else to be done, and some choose to put down terminal animals themselves using a gun. The best spot to shoot a cow for a humane death is in the center of the forehead, about an inch above the space between its eyes.

If you euthanize or an animal dies on your farm you will have to dispose of the corpse. There are services available that pick up the carcass. If you do not know what caused the illness, call a university diagnostic lab to run tests. Many university labs will do tests for free, but you may have to transport the dead animal to their offices. Or, bury the corpse. Check with county or state officials about laws regarding burial. Usually, you must keep the corpse a certain distance from natural water sources, for example 100 feet, and a certain distance from other homes, for example ¼ mile. Your area may also have laws about the depth to bury the carcass — 3 or 6 feet, for example — and the entire animal should be covered with dirt.

If you have a large farm, you could also leave the corpse out in an isolated area for birds or other scavengers. You may not be able to do this is if the animal was euthanized by a veterinarian because the poison could harm animals that eat the contaminated tissue. This may also be the case if the animal has certain contagious diseases such as foot-and-mouth disease, a viral disease that causes blisters in the mouth or on feet that make it uncomfortable to walk or eat. If you have questions about whether it is safe to bury an animal on your farm, ask your veterinarian. Your veterinarian or county and state officials should be able to tell you all the rules regarding on-farm disposal.

A good health plan ensures you only have to euthanize a very small percentage of your animals. Vaccinations, prompt treatments, and clean, natural living conditions will mean healthy lives for the animals in your herd. *In Chapter 7, you will learn how to feed them in times when there is little good pasture to be found.*

• Chapter 7 •

Preparing for Nature's Extremes

TERMS TO KNOW:

Relative Feed Value (RFV): This is calculated by the amount of energy and protein present in the hay to determine its nutritional content in relation to grass.

Drought: A prolonged period of dry weather

Sacrifice area: A place where you keep your cattle for extended periods of time when the ground is vulnerable, such as winter or during heavy rain

Mother Nature is a cruel mistress. There will be years when everything goes well, and it will be easy to get the most out of your herd. And then, there will be years when you will have to deal with droughts or heavy snows that make it tough to meet the nutritional needs of your animals. You may have to make tough decisions about buying extra hay, which adds to your farm expenses. You may even have to sell off animals because you will not have enough pasture.

When making these decisions, your main priorities should be protecting your pasture and protecting your breeding animals. Cow-calf farmers who make their living by producing calves every year depend on their mother cows and their bulls to perpetuate their businesses. If you do not protect your pasture, you will not have the resources to feed your herd when the weather does improve. The good news is if you expect the unexpected, you can decrease the damage wrought by nature's extremes.

For much of the country, there will come times when farmers have to deal with one extreme or another. In the northeast and upper Midwest, harsh winters can shorten the growing season to less than half the year. In parts of northern Minnesota, the growing season is less than four months. In parts of the West, annual rainfall can be fewer than 10 inches, and the growing season can be halted by months of scorching heat.

The good news is there are grass-fed cattle farmers all over the United States — proof that these challenges can be managed. One way you can reduce costs during the dormant seasons is to grow various plants to extend the grazing season. Each extra day you can add to your growing season is an extra day your herd can get its nutrition from the land. It is also another day you do not have to provide expensive stored feed. This is important because stored feeding is one of the highest expenses on farms. *In this chapter, you will learn about these tricks and about managing your stored reserves to ensure your cattle get the nutrition they need for your beef or dairy operation to thrive.* You will also learn how to plan for unforeseen events such as droughts or especially harsh winters. First, a look at some of the challenges nature may throw at you:

Drought

A drought is simply a prolonged period without rainfall. Droughts are tough because they can sneak up on you, and there is no way to tell how long it will last. Without this precipitation, your pastures will dry up, stop growing, and turn brown.

You may have to supplement your pasture with hay during times of drought. This is expensive but manageable if your shortfall is not too bad or you do not have too many animals to supplement.

The danger here is you could run out of pasture grasses or stored forages. You can decrease the damage if you have a plan in place. Some farmers stubbornly wait out bad weather without making any changes, which means they either do damage to their pastures by overgazing, they run out of pasture and have to sell more animals than they would have if they had reacted sooner, or they pay too much for hay because they buy when demand is at its highest. The earlier you react to signs of drought, the better chance you have of buying hay before increased demand drives up the price, which means you will beat other panicked sellers to the stockyards before they drive the price up.

Winter Grazing

A common challenge of winter grazing is many grasses are dormant. As discussed in Chapter 2, plant different types of plants that will grow when other pastures are dormant, such as corn, rye, brassicas (rape and kale), and chicory. To spread the nutritional value of these plants out over the largest number of days, strip graze these plants by using an electric fence to ration in sections.

In some areas of the country, snow makes it hard for cattle to graze. Plants that grow tall, including corn in its vegetative stage, can help combat this problem because they are more visible through snow than shorter grasses, which can be buried in white powder. Cattle can graze in snow and will learn to dig into light layers to look for grass.

Another problem with winter grazing in many areas of the country is the ground is often wet and vulnerable to damage from hooves. This is also true at other times of the year after heavy rains. Some farmers account for this vulnerability by using a sacrifice area. A **sacrifice area** is a place you keep your cattle for extended periods at times when the ground is vulnerable. The sacrifice area gets torn up from heavy use and will not grow good grass in the spring, but the rest of the pasture is protected from damage. When cattle are on a sacrifice area, you will probably have to feed them hay because there will not be enough grazing grasses to sustain your herd for long.

Growing and Storing Food Reserves

In parts of the country where grasses do not grow year-round, have a good reserve of forage to provide food on days when your pasture cannot support your cattle. The most common stored forage is hay. Hay is dried, palatable forage. It is usually collected into small square or large round bales, but it can also be raked into rows in the pasture and grazed there. Hay can be made from many different plants, but some common ones are clover, oats, and millet.

Large bales range from 800 pounds to as much as 1,800 pounds. Cow size and bale weight should help you determine how long a bale will last or how many you may need for a certain amount of time. Farmers usually estimate each head of cattle needs about 30 pounds of hay each day they cannot graze. Before your pasture yield improves from your rotational grazing system, you may need at least 120 days of storage forage, possibly much more in regions of the country with heavy snowfall or little annual rainfall. If you have 30 animals that need hay for 120 days, that means you need 108,000 pounds of hay. If you use 800-pound hay bales, then you need 135 bales of hay.

Many farmers reserve parts of their pastures for use as hay fields where they harvest hay or other forage-based reserves. If you do not already have the equipment to make hay, contract with someone who does have the equipment and is willing to cut and bale the hay for you. It might cost $11 or so per 1,000-pound bale to get someone to cut and bale for you, and this could quickly add up to more than $1,000 for a 10-acre field. A more common arrangement is to have someone with equipment come to your farm to cut and bale it on shares, which means you do not have to pay money but the contractor gets to keep half your hay. These people will advertise in local classifieds and probably also will be well known among farmers in your community.

If you do not have land reserved for hay fields, buy your hay from someone who is selling it. Some sellers advertise their hay in local papers, or you may even notice signs by the side of the road. A website that lists sellers by state is The Internet Hay Exchange (**www.hayexchange.com**). Another site you may find useful is The Hay Barn (**www.haybarn.com**). This site has classified ads from hay sellers and buyers. The cost of hay depends on supply and demand. In times when hay is readily available, you may be able to buy a 1,000-pound bale for around $25. In times when hay is scarce, such as in a drought, a 1,000-pound bale could cost $60.

Hay has to dry out before it is baled. Freshly cut hay is usually high in moisture, and if it is baled, it begins to ferment. If there is a normal amount of moisture, the hay heats up but eventually dries out. If the hay has been oversaturated, such as hay that gets rained on before it is baled, it will rot. Hay that was rained on but still has not dried out enough before it is baled can undergo a chemical reaction that causes it to combust. Another way to preserve cut pastures is to bale it while it is still high in moisture and wrap it in plastic. This allows it to ferment but keeps out the oxygen necessary for combustion. These high-moisture bales are called baleage or haylage. *They will be discussed later in the chapter.*

Many farmers will designate a certain spot in their field where they will keep their hay and simply move the bales to there. If you leave your bales in the field, bigger bales are best because the outer layer of the hay will protect the inside from spoiling. You can also cover hay with a plastic sheet or a tarp to protect it from the elements. Hay bales stacked outside must be covered with these sheets. Store hay in areas of the field not prone to flooding or standing water. Round bales can be stored in rows, end to end, but with space in between the rows. If you store bales inside, elevate bales off the floor by setting them on wooden pallets. You can also stack bales on top of each other in the field If you do this, you must cover the tops with a tarp, or rain will get in between them and cause them to spoil.

Many farmers will feed their hay to cattle in various spots in the field to get the same benefits as rotational grazing. By putting bales in different paddocks and moving the cattle around, you will continue to distribute manure across the field and reduce the damage done by animals lingering in one area of pasture. Some use mobile feeders. These come in various styles, similar to the mobile troughs that can be pulled around the field. You can also use a skid steer to move bales to areas of the pasture where you want your animals to graze.

If you let cows have unrestricted access to hay bales, they will waste much of it by using it as bedding. Restrict access to the bales with hay rings, metal storage units that encircle bales and prevent cows from bedding there. You can also ration bales with electric fences in the same way you might strip graze, by moving the fence back to reveal one bale at a time.

Not all types of hay have the same nutritional values. Hay is classified according to relative feed value (RFV). RFV is calculated by the amount of energy and protein present in the hay. To maximize RFV, forage must be harvested at optimal maturity, which varies depending on the plant. Forage guides published by universities or state Natural Resource Conservation offices usually include the best time to harvest plants. The goal is to find the point when you have the best nutrition and the most vegetation. For most grasses, this point is after the seed head has formed but before the seed has matured. Alfalfa harvested in this stage is considered the best plant to use for hay. Like other plants, if alfalfa is harvested in later stages of maturity, it has a low RFV. Take samples to your local extension office to test for RFV.

Hay is not the only type of stored forage. Other options are:

Silage: Silage is fermented, moist forage. It is stored either in silos or in concrete bunkers covered with air-tight tarps. If it is not stored properly, a dangerous bacterium called *Listeria* could sicken or kill your animals. Silage can be made from almost any green, growing plant. Corn silage is the most common form and is acceptable in grass-fed production if it is cut while in the vegetative stage, before it begins forming the cob. Due to its high moisture content, silage is typically only used on the farm where it is made. If you do have a nearby source, it makes a nutritious, highly palatable food for beef cattle. It must be fermented and stored properly or *Listeria* can become a problem.

Haylage: Haylage is a stored forage cut and baled like hay but cut and stored to preserve its moisture content. The bales are immediately wrapped in airtight plastic to prevent them from drying out. This plastic also keeps out the oxygen that would cause combustion in hay. Haylage can be stored outside. At 45 to 55 percent moisture, haylage is drier than silage but contains more moisture than hay.

Baleage: Baleage looks like large, round hay bales but is baled at a higher moisture percentage and allowed to ferment. Bales are wrapped in multiple layers of plastic.

Pasture strategies

There are also a couple of strategies you can try to get the most out of your regular pasture grasses, including stockpiling sections of pastures and swath grazing. Discussions of each of these strategies follow.

Swath grazing: Swath grazing is an alternative to baling hay. In this strategy, grasses are cut the same way as hay at the end of the grazing season, but they are not baled and instead are left where they were cut or raked into rows. This eliminates the cost and effort of baling and allows cattle to graze in a manner similar to when the grass is growing. Cutting for swath grazing can also be done with small grains in the vegetative stage. Use electric fences to give cattle access to one or two swaths at a time, similar to the way you can strip graze pasture grasses.

Stockpiling: Another strategy farmers employ to extend the grazing season is to stockpile sections of pasture for use in the dormant season. Stockpiling simply means keeping cattle off certain areas so they can come back and graze it after all the other parts of pasture have been grazed and are not growing. Stockpiling is accomplished by not letting cattle graze an area for the last 70 or so days of the grazing season. Usually, this means cutting it for hay in July and then leaving it alone until late October. A commonly stockpiled plant is tall fescue, which does not lose its nutritional value during the stockpiling period. Although the toxic fungus endophyte can make tall fescue unpalatable in summer, the toxicity levels are reduced when stockpiled fescue is grazed later in the year. Strip graze stockpiled forage to balance out the nutritional quality over a larger number of days.

Many of the farmers interviewed for this book own or lease additional pastures to give them additional grazing land for winter. Some of them were fortunate enough to be able to rent pastures adjacent to their own land; others rent pastures within a few miles. Cattle can be moved in trailers to these off-farm pastures in winter and moved back in spring.

When Reserves are Low

In times of extreme weather such as droughts or harsh winters, you may find you did not budget pasture or supplements to feed your herd for the rest of the year. In this case, you must act to make sure you have enough food to feed your animals without overgrazing your fields. As discussed in Chapter 4, if you overgraze your grass, you diminish the stored energy in the roots, which weakens the plants and their ability to recover. In times of drought, one option you have is to feed your animals hay. This allows you keep rotating your animals through the pasture without putting pressure on your grass reserves. When grass starts growing again after rain returns and the weather cools, your pastures will be in better shape to recover. But in drought situations, the cost of hay also increases, which means it is more expensive to keep your animals in good shape. If there is not enough food to go around, your animals will not be able to grow to their full potential or to provide as much milk as they otherwise could.

Reducing cattle stock

Another option you have is to reduce your herd size. **Destocking**, or culling cows, will reduce the pressure on your pastures and will ensure you will have enough grass when rain returns. When you destock, target animals you would have sold first anyway, such as cull cows and stockers. The cull cows include heifers that did not breed and steers you were raising as stockers. If the drought worsens, you may have to cut deeper into your herd than you had planned. For example, you may have to take animals you had wanted to raise to finishing and instead auction them off at a sales barn. Although this will mean a financial hit, especially if everyone in your area is reducing stock and market prices dip, it will be better in the long run than overgrazing your land. The goal here is to preserve your breeding stock. If you have to sell cows that breed every year, then you are really taking a financial hit not only this year, but every year in the future in which that cow would have provided you a calf.

If you wait too long, you may have to sell some of your breeding stock and cost yourself even more down the line. If you sell before everyone else begins to, you can alleviate some of the damage because prices will continue to drop as the weather crisis worsens. Balance this by slowing down your grazing and supplement with hay or other non-grain supplements. Another approach is to sell more weaned calves than you had planned so as to reduce the pressure on your pasture. Have a plan if you need to reduce stock further. For example, sell the oldest cows or bulls first. This can be heartbreaking if you have an old bull or cow that has been good for you over the years, but it is a safer bet the younger animals have more productive years left in them, and each year you keep them around is a year they are making money for you. Farmers are often forced to keep ornery cows around for breeding because there are not suitable replacements, but target these animals for sale if you need to reduce stock. Also, sell cows whose calves do not grass finish as easily as some of the others. *In the next chapter, we will talk about the life span of cattle.*

CASE STUDY: ROCKY MEADOW FARM

W.R. LeClair, owner
Rocky Meadow Farm
Francestown, New Hampshire
Contact Information:
wrleclair@myfairpoint.net

W. R. LeClair's 270-acre farm was almost a blank canvas when he bought it in 1985. Trees had claimed land that had once been a farm. So he cleared an area for the house and barn and started clearing the original fields by cutting timber, digging up rocks, and removing stumps. He now has 50 acres cleared, and chips away at more each year.

His pasture grasses come from a seed mixture by the Barenburg seed company and include perennial rye, white and red clovers, tall fescues, orchardgrass, and timothy. Not all seeds from the mix thrive in each microclimate of his farm, so some grasses grow better in some spots than others. He does not plant it, but wild Kentucky bluegrass sprouts naturally in places. He fertilizes most of his fields once a year, and soil tests sometimes spur him to add nutrients such as boron and potassium.

LeClair's fields are all small, 5 or 10 acres, and they all flow into each other. He uses about six fields to rotate his cattle through and the rest for hay, which he needs a lot of in the Northeastern winter. Every field has its own separate water supply: ponds, free-flowing streams, or frost-free waterers from his wells (the animals learn to push down the ball on top of the waterer to drink from the reservoir). His cattle recognize him as a food source, so he has no problem moving them from one field to another.

"I move them by myself like the pied piper," he said. "I just show up with my Kawasaki mule, and they know it is time to move. Just by habit, instinct, and experience, they just follow the mule. I just take them from one field to the next."

LeClair, a physician, chose grass-fed cattle because of the health benefits of grass-finished beef. His breed of choice is the solid-colored Galloway. He discovered them by accident in an article about heritage breeds, breeds that used to be common in certain areas they were adapted to but became threatened by industrial agriculture's focus on a narrow number of breeds. LeClair was attracted to the characteristics of solid-colored Galloways listed in the article — the Scottish breed is adjusted to cool, damp climates, and the animals grow well on grass.

"I found that to be a fairly rare breed," he said. "It piqued my interest because of its advantages, specifically that it does well on grass only, it has a mellow personality, and it is very hardy. They are kind of interesting animals to look at; they are attractive, I think, as cows go. It hadn't been popularized yet as much as the Hereford or the Angus."

Most of LeClair's older mother cows came from Montana and a few came from Vermont. His first bull came from Alberta in 1998. He usually slaughters about 20 animals per year, and he also sells some heifers and bull calves as breeding stock to farmers in New Hampshire, Vermont, New York, Massachusetts, and Pennsylvania. He is considering getting a Devon bull to cross with his Galloways.

"The Devon's got a very sweet, tender beef flavor," he said. "They tend to put on a lot of meat. They fill out pretty well. It might give me a little bit of crossbreeding vigor I wouldn't get if I kept inline breeding."

In winter, from late November on, he feeds his animals baleage he cut from his hay fields in the summer. He feeds them in round bale feeders or from a big wagon with eating stanchions on the sides. He can feed 18 cows at one time, so four 1,200 pound bales will feed them for a week during winter or if he needs to go out of town for a week. He also supplements pasture with baleage in the hottest part of summer. His big trailer is mobile, but he tends to keep it in one spot in a 5-acre field lower than the other fields. This lower elevation provides shelter from cold wind. He makes about 200 bales per year. He uses about 75 and sells the extras.

"It is naturally protective and an excellent spot for overwintering these cattle," LeClair said. "It just does not get the wind. They do not go in the barn, not at all. It is actually healthier for them to be outside."

He sells his beef two ways, either by the side for $4.35 per pound hanging weight, or in a 30-pound variety pack for $235. He does not sell individual cuts because he thinks it takes too much time.

It usually takes him about 24 months to finish an animal, sometimes 26, and 1,200 pounds is about the right finishing weight.

"You are always trying to get more fat on the animal, and sometimes, it works very well," he said. "But sometimes, you are leaner than you would like. The Galloway is naturally lean anyway.

"There is a long learning curve in grass finishing beef. I think there are no absolute experts. I think everyone is learning all the time."

• Chapter 8 •
Life Cycles

TERMS TO KNOW:

Nurse cow: A cow that nurses the calves of other mothers as well as her own calf

Grafting: The process of introducing calves to a nurse cow

Seasonal calving: Putting bulls and breeding-age females together for a predetermined period of time to ensure all females will get pregnant and deliver calves in the same time window. Calving seasons usually last 60 days or 90 days. Calving seasons are targeted for spring or fall, though summer is not unusual. Some farmers have two calving seasons: one in spring and one in fall.

Seasonal dairy: A dairy that delivers all its calves in one season. These dairies stop milking for the last couple of months before the calves are due.

Farmers play an important role in the life cycles of their cattle. They determine when the animals will be born, the conditions in which they live, and, in most cases, when their animals will die.

The lives of your cattle also will be linked to the seasons of the grass. As discussed in Chapter 6, cattle have high energy requirements at calving and during lactation. Beef cattle will need to start their lives at times of high forage quality and spend the last weeks of their lives eating similar high-nutrition vegetation. With the possible exceptions of the few areas of the country where good pasture is available year-round, you will have to manage your herd so their life cycles are in sync with your pasture's growing cycle.

How Various Calving Seasons Will Affect Your Farm

In general, a cow's pregnancy lasts 280 days. Cows are re-bred about three months after they deliver, and calves are weaned at about 6 months old, though they could be older or younger.

Most farmers try to breed all their cows in a certain window to avoid inclement weather and to reduce the amount of time spent providing labor for the breeding and birthing processes. They try to impregnate all their females within a 60-day or 90-day window so all the calves will also be born within 60 days or 90 days of each other. Some farmers try for an even shorter window of 42 days. Unless you are using artificial insemination, you breed your cows by turning your bull loose with them. Remove him at the end of the breeding window.

Some farms, especially in the southeast or in California, have much longer breeding seasons because there is little or no cold weather to avoid. Some farmers will simply leave their bull with the herd most of the year. You can also leave the bull with the herd if all the heifers are bred and keep him there until the new heifers reach puberty. This allows farmers to stagger their calving so they will have calves reaching selling age at various times of year, which provides income throughout the year rather than all at once, as is often the case with seasonal calving.

Spring

Many beef farmers want calves born in the spring to sell in the fall after weaning. A spring calving season is also common for seasonal dairies. In an early-spring calving system, cows are bred to deliver their calves in late February or early March. The benefit of this system is grass will begin growing about the same time the calves are born and will provide the extra nutrition needed for the lactating mothers. A disadvantage of early spring calving is if calves are born too early in the year, the weather might be poor and the calves could arrive in snow or mud. A muddy or wet calf can get chilled and die. The weather may also be unfavorable during the breeding season, beginning in about June and carrying into July, when hot weather could reduce fertility. A way to avoid this is to target calving for late spring, such as April or early May. Late-spring calving allows cows and heifers in late-term pregnancies to get the additional energy they need from the fast-growing grass. This also ensures heifers that are still growing get the energy they need to improve the chances they will re-breed for next year. By allowing the pregnant cows to get the spring's first growth, you also reduce supplemental feed costs. Get an idea of the first day of spring by using the National Climactic Data Center's searchable database of temperature and precipitation data, which includes charts that provide the probable dates of first and last freezes in each state and in different areas of each state (**http://cdo.ncdc.noaa.gov/cgi-bin/climatenormals/climatenormals.pl**).

Summer

Summer calving dates are usually between early June and the end of August. Summer calving works best in areas where warm-season grasses are predominant, such as Nebraska and the Northern Plains. In these areas, grasses are still growing in summer, which allows cows near birth to get the big energy intake needed to produce calves and milk. In summer calving, births happen in warm conditions that are less dangerous for the calves than the cold, wet surroundings that accompany early-spring calving. Summer calving also gives you time to wean calves before grass grows dormant in winter.

Fall

In a fall calving system, calf delivery is targeted for September through November after grass has started growing again and following the hottest weather of the year. Fall calving systems avoid the problems of births during cold, muddy spring. Time your breeding to avoid the heat or worst winter weather. Seasonal dairies in the hottest parts of the country benefit from fall calving seasons because they avoid the brutal heat of summer when grasses will not grow. These parts of the country have more grass available in winter than colder areas, so farmers in these areas do not need to provide as many feed supplements for lactating mothers. Weaning occurs in the spring, so the calves can graze fresh grass. Fall calving also allows weaned calves to hit the market in the spring when prices usually are higher. If prices are low in the spring, hang on to your cattle through the summer and hope the prices will increase.

Two calving seasons

Some farmers have two calving seasons to better distribute income throughout the year. Dual calving seasons are attractive to some farmers who might otherwise need more than one bull. By dividing the number of cows to be bred into two groups, you can service more cows with one bull. For example, if you have 60 to 70 females to breed in a single calving season, you would probably need two bulls. If you bred half of those 60 to 70 in the spring and half in the fall, one bull could do the job. The downside to two calving seasons is more management is required as you will have to keep your two groups of females separate during breeding.

If you use more than one bull, put them together away from the herd for at least 30 days before you try to use them for breeding. This time together will allow them to establish their pecking order. If they have not yet determined which bull is dominant before breeding season, they will fight each other for dominance, possibly even knocking each other off the females during mating. Such battles increase the likelihood of injuries, and a bull with an injured penis is useless.

Seasonal dairies

Many grass-based dairies are seasonal, which means their calves are all born at the same time of year and the milking cows are all dried off at the same time so there is a period of time when milk is not produced on the farm. The energy that would have been used for milk production will instead be used for the fetus. The cow may be uncomfortable soon after you stop milking her, but she will stop producing milk. Milk production decreases late in the milking season naturally.

Your dry period will depend on the timing of your calving season. Most farms time their dry periods to coincide with times when grass is dormant because cows that are not lactating have fewer nutritional requirements than those that are lactating. In most areas of the country, this dry period will be the winter months. In warmer climates, such as the southeast, winter grazing can be strong, and you can time your dry period for the miserable summer months.

Breeding

Many farmers like to try breeding their heifers 30 days before cows that have already produced calves. Most cows will be re-bred 90 days after the first calf is born while the first-calf heifers would be re-bred 120 days after their first calf is born. This extra 30 days gives the younger females an extra breeding cycle to recover. Females that have only delivered one calf are often referred to as first-calf heifers. When it is time to re-breed first-calf heifers the next year, breed them with the rest of the herd.

At breeding time, which should be when she is about 14 months old, a heifer should weigh at least 65 to 70 percent of her mature body weight. At calving time, she needs to weigh at least 85 percent of her mature body weight.

One way to determine if a heifer is suitable for breeding is to measure her pelvic area to ensure the opening can accommodate the expected size of her calf, which can be estimated based on EPD data. *This is discussed in Chapter 5.* A veterinarian can help you measure the canals of the heifers on your farm. If you are buying a heifer from a breeder, he or she should be able to provide you with this data.

As discussed in Chapter 5, the most reliable sign of a cow in heat is she will allow other cows to mount her. Observe the animals twice a day, in the morning and in the evening, to observe this behavior. Secondary signs of heat include pacing or restlessness, sniffing, and nuzzling, or a string of clear mucus hanging from the vulva or smeared on the hind legs or tail. If you detect a cow in heat, try to breed her within 12 hours because heat lasts between 12 and 18 hours. The cycle from one period of heat to the next is called the **estrous cycle**. This cycle usually lasts about 21 days.

Confirm pregnancy with pregnancy tests, which usually involve a veterinarian performing a rectal palpation of the cow's uterus and ovaries. Another indication of a pregnancy is failure to return to heat when a cow's next heat cycle is due. Dairy cattle are generally checked for pregnancy between 28 and 35 days after being bred; cows that are not pregnant can be re-bred.

Most cow-calf farmers and dairy farmers want their cows to give birth every 12 months. To accomplish this, cows must be re-bred three months after the first calf is born.

This waiting period gives their reproductive tracts time to cleanse themselves. They can still nurse their calves after they are re-bred, and dairy cows can still be milked. Breeding cows can stay productive into their teens. On conventional dairy farms, breeding cows may only live long enough to give birth to a couple of calves.

Shifting the calving season

If you are currently farming and think a different calving season would work better for you, it is possible to shift your calving season. If you want to move from fall to summer or spring, you have to start breeding a couple of weeks earlier than you did the year before, and then start a couple of weeks earlier the next year, and so on, until you reach your target schedule. The challenge of this is not all your cows will have had much time to recover from the previous calving, and some may still be in a period of infertility following childbirth. If you can afford to cull the cows that take longer to conceive, you will ensure you will be working with the most fertile animals in your herd.

One trick is to synchronize your cows' cycles. Conventional farmers sometimes give their females a hormone shot to trigger estrus, when the female will be most receptive for breeding. However, hormones are banned under organic and AGA rules. If you feel it would be beneficial to synchronize their cycles, there is a natural way to accomplish this: Remove the calves from the mothers for 24 to 48 hours. This is stressful to both mothers and calves, and it is actually the stress of an apparently lost calf that triggers the mothers to begin cycling. Be careful when trying to separate the mothers from their calves; they can be very protective. Bring the whole herd into your handling facilities and separate them into different pens, or try to chase the mothers away so you can get the calves into a trailer. This stress will also cause your calves to lose weight, which is not ideal for beef farmers because you want calves to be gaining weight their whole lives. You will have to determine if the risks outweigh the benefits to your farm.

If you wish to shorten your calving season into a smaller window, remove the bull earlier than you have in the past. This gives your cows a shorter window in which to conceive.

From Birth to Pregnancy: What to Expect

The last 90 days of pregnancy are critical to fetal development. It is during these last few months of pregnancy that a cow's body shifts its energy intake toward the fetus. In dairy cows, this means that milk production slows noticeably. To aid in fetal development, feed your cows an extremely high-protein diet, which will usually be satisfied through pasture grasses though you may have to supplement in a drought or winter. If you are calving in the spring, be sure your mineral mix is high in magnesium to prevent grass tetany. *This is discussed in detail in Chapter 6.*

Birth

You will be able to notice changes in a cow as she nears her delivery date. Her udder will start to become firm and full about ten days prior to calving. The vulva and tail head may become swollen and loose and jiggle when she walks (this is called **springing**). She might stand with her tail raised, possibly in isolation from the rest of the herd.

Most births will be uneventful. Calves will be born out in the pasture without assistance. But when females are ready to calve, it is important to be in clean, dry pasture. If it looks like a calf is going to be born in cold weather, you might need to move the mother indoors to protect the calf. The calf must be sheltered from the wind and rain. If you do not have a barn, use trees, hills, or windbreaks to protect the calf. If the mother appears to be having difficulty, you and possibly a veterinarian should assist in the birth. If you need to assist with a birth, you may need the following tools:

- **Paper towels**: These will be nice to have to keep your hands clean.
- **Colostrum**: This is the nutrient-rich first milk calves must take to jump-start their immune systems. If the calf does not begin suckling, you will have to feed it yourself. Many dairy farmers collect colostrum from their cows within 24 hours or so after births and keep a supply on hand. You can also buy it in powdered-milk form at farm supply stores. One store-bought bag equals one dose, and these bags will have instructions on preparing this mixture for the calf. *Colostrum will be discussed in further detail later in this chapter.*
- **Bottle or feeding tube**: These will be necessary to give the calf colostrums. You can try to bottle-feed the calf, but it will probably be more practical to force-feed the calf using the tube.
- **Blanket or calf warmer**: These are specially made blankets used to keep calves warm. Blankets might be necessary for sickly calves or calves born in winter in colder states. They can be found at farm supply stores or online at stores such as Calf Cozy (**www.calfcozy.com**).

- **OBGYN chains and handle or a calf jack**: If you need to assist with the birth, you may have to pull on the calf to help it out. These chains loop around the calf's leg and the handle attaches to the chain to give you something to grab for pulling. The chains and handle will cost less than $20 at most farm supply stores. Simple ropes could also be used to loop around the calf's legs. A more sophisticated method for pulling calves is a calf jack. A **calf jack** involves looping a chain around the calf's legs and using a hydraulic handle like you would see on a car jack to pull the calf out, 1 or 2 inches at a time.

- **Gloves**: Gloves may be hard to work in, but you might be glad to have a pair of shoulder-length gloves if you have to reach into the birth canal to adjust a calf during a difficult birth.

- **Soap**: Many veterinarians would recommend washing your hands or cleaning the birth canal before you reach in. Regular bar soap would be fine to use in this instance.

Many calves are born without assistance. Farmers simply will come outside to check on their cows and discover an addition to the herd. But occasionally, for reasons such as cold weather in a spring calving scenario or because of calves in awkward positions as they exit the birth canal, cows need to be assisted with a birth. Sometimes, this assistance is relatively simple and easy to manage. Other times, a veterinarian will need to perform a caesarian section to retrieve the calf. As you get experience, you will learn how to help and when you will need help from a professional. Early on, if you have doubts about how to help, it is a good idea to call a veterinarian for help.

Pregnancy occurs in four stages:

Stage one: During this stage of labor, the cow will experience contractions in the uterus. These will make her become restless, and she may get up and down frequently, swish her tail, or kick at her sides (this will not hurt the calf).

During this stage, her body is preparing itself for the passage of the calf; the ligaments relax, and the cervix, vagina, and vulva are dilating. This stage of labor can last as little as three hours for cows, but heifers can remain in this stage for up to three days.

Stage two: Stage two of labor begins when the water bag appears at the vulva and exits the body. The calf has moved into the birth canal, and the mother will start to strain and push. This stage will last about 30 minutes for cows to up to three hours for heifers. You can probably leave her alone during this phase, but if it takes more than two hours for the calf to appear at the vulva, you will need to assist in the delivery. Occasionally, the cervix will fail to dilate wide enough for the calf to pass through. If this is the case, call your vet. When checking, veterinarians recommend you use soap and water to wash the vulva with your hand and arm.

Wear a rubber glove or an obstetrical sleeve, and then insert your hand into the vagina. The cervix may not be dilating properly if it has only dilated the width of two or three of your fingers. Another possibility is the uterus may be twisted. If you think this could be the case, call your veterinarian immediately. When the cervix is fully dilated, you will be able to feel two front feet and a head or two rear feet and a tail at the opening of the cervix or in the vagina. In a normal birth, the front hooves will appear first, with the calf's head on top of the legs, almost in the position of someone sliding on his or her stomach.

If the cervix is dilated properly but the calf still is not coming, this could mean uterine inertia or poor uterine contractions, and you will have to assist with the birth. Use chains or knotted ropes and slip one rope or chain above the lower joint of the legs and one rope or chain below the lower joint of the legs. Time your pulls with the cow's pushes. If she has not been pushing, she will start when you begin pulling. Pull in a downward motion. It should only take two or three pulls. If you still cannot get the calf out, call your veterinarian.

Stage three: Stage three is the stage when the calf actually comes out. In a smooth delivery, delivery has reached this stage when the calf appears at the vulva. Sometimes the calf will come out hind-feet first; this is called a breech birth. Do not be too concerned as long as both back feet are coming at the same time, but in a breech birth you will probably have to pull. You must also help or call for help when the calf comes out in other strange positions, called dystocia. If it is something simple such as a foot being flexed back, correct this simply by pulling the foot into the normal position. If the position is more complex, such as legs bent back, the head twisted to the side, or the calf lying sideways across the birth canal, you will probably need to call a veterinarian for help so you can protect the uterus from tears and to protect the calf.

Stage four: In stage four, the placenta passes. The cow might eat the placenta; this is normal and natural. If she does not, bury it so it does not attract buzzards. It may take a few hours or even a few days for the placenta to pass. It usually will not come until after the mother has cleaned the calf and the calf has begun suckling. If you have had to assist with the birth, check the uterus to be sure there is not a second calf by sticking your hand into the birth canal. If the labor was difficult, also check the uterus and vagina for tears or excessive bleeding. In a smooth birth, there will be no uterine tears. Call your veterinarian if you find any tears or if the uterus is **prolapsed**, meaning on the outside of the birth canal.

A Calf's First Hours

The first hours of a calf's life are very important. Immediately after delivery, the calf should be blinking and attempting to lift its head. Check to make sure all mucus is removed from the mouth and nostrils. If it is not breathing, tickle its nose with a piece of straw to stimulate breathing. If that does not work, try chest compressions or even blow into its snout to force it to take a breath. If a calf is gurgling fluid, help the fluid escape from the respiratory tract by hanging the lower half of the body over the side of a pen or two bales of hay or straw so the head hangs down below the upper body. Chest compressions while the calf is draped can also help with breathing.

The cow needs to clean and dry off the calf, so either get the cow up or gently drag the calf to her head so she can begin. The umbilical cord usually breaks during birth, so you will not need to cut it. If the birth has been smooth, the best thing you can do is get the mother and calf together and then get out of the way. If the calf is suckling and appears healthy, you do not need to do anything extra.

It is also important that the calf get its necessary colostrum. Colostrum allows the mother to pass on her antibodies to her calf, which is born without them. These antibodies boost the calf's immune system and provide the protection it needs for a good start in life. If the calf begins suckling, it will get enough. The calf should start suckling within two hours. If you find a calf born unassisted, keep an eye on it to see if it is nursing normally. If a calf is not suckling, feed it colostrum. If a calf does not get colostrum, it will not be able to absorb the crucial immunoglobulins it needs to fight infections. Calves that do not get this colostrum are more likely to get sick and die. Sometimes, a mother will abandon her calf and will not let it feed from her. If this happens, or if the calf will not nurse, bottle-feed colostrum or force-feed it with an esophageal tube. *Feeding electrolytes to sick calves is discussed in Chapter 6.* Force-feeding may actually be the better option to ensure it gets the amount it needs quickly. Many farmers keep a frozen supply of colostrum on hand. They let calves get first dibs on mothers' milk, but within 24 hours or so they collect some to keep on hand. If you do this, do not overly heat the colostrum because you may kill some of the beneficial content. Just let it thaw in warm water.

Nursing

Beef cows are usually left to nurse anywhere from six to nine months. Dairy calves are handled differently. *This will be discussed a bit later in the chapter.* A newborn calf does not have a fully functioning rumen. When a calf drinks milk, folds of tissue make a groove from the esophagus to the last stomach, the abomasum, bypassing the rumen. It takes about four months to develop a fully functioning rumen. Help the development by introducing roughage such as grass or hay at an early age. Calves in pastures will begin to graze even as they are in the nursing stage.

Dairy calves

In conventional dairies, young calves are separated from their mothers and bottle-fed a milk replacer substance, which often contains antibiotics and other substances banned in organic or AGA systems. Whole milk from other cows on the farm is sometimes used instead. The goal of grass-fed dairies is to allow their animals to live as naturally as possible, and this usually includes letting the cows drink real milk. Living naturally means coming up with creative ways to feed calves while still producing enough milk to sell.

One option for feeding your calves is mob feeding. **Mob feeding** involves keeping groups of four to ten calves in their own paddock separate from their mothers. Instead of nursing, the calves are allowed to eat from a mob feeder, a big barrel with nipples on it. The mob feeder will contain milk diverted from the milking herd. This is better than milk replacer, but some farmers feel mob-fed calves do not grow as well as calves allowed to nurse.

Using nurse cows is another option farmers have used for feeding dairy calves. **Nurse cows** are cows that will not only nurse their own calves but also the calves of other cows at the same time. You might be able to get a nurse cow to feed three or four calves at a time. At the Yegerlehner farm in Clay City, Indiana, potential nurse cows were initially selected because the family did not want to milk them with the rest of the herd. The first candidates had mastitis or were cranky during milking. Alan Yegerlehner, who owns the farm with his wife, Mary, said a good nurse cow must be willing to accept other cows' calves. Even if she kicks at them during the initial feeding, with persistence she can be trained to accept them. If she keeps kicking after two or three days, it is better to try another cow. The Yegerlehners use some cows as nurses every year because the cows simply show an instinct for that role. A nurse cow must also have enough body fat reserved to support the intensity of milking several calves. You can tell by looking at cows if they are fat enough.

Uniting calves with a nurse cow is called **grafting**. Before grafting, calves can be left on their mothers for at least a day or more so the calf gets a good start on its mother's colostrum. Graft all the new calves onto a nurse at the same time. Keep the calves separated from their mothers and kept together; a good place to keep them would be your holding pen. Letting the calves fast for about 24 hours will ensure they will be especially hungry once they are introduced to the nurse cow.

During the grafting process, a potential nurse cow is caught in a head gate so she cannot try to avoid the calves that are not hers. After the first feeding, calves are separated from the nurse cow and put into their pen until the next feeding. It usually takes four to eight feedings over the course of at least a couple of days to complete the grafting process. After a couple of feedings, try one feeding without restraining the nurse cow in a head gate. If everything goes OK, the calves and the nurse cow are left together to bond for a couple of days before they are returned to the pasture. From then on, each grafted calf must be responsible for its own feeding because the nurse cow will not find them to make sure they eat. Nurse cows and calves are usually kept as a separate group from the milking herd because if both groups were kept together, calves would return to nurse from their birth mothers.

The nurse cow system is not perfect. The energy required to nurse multiple calves takes a toll on a cow, and a potential side effect is nurse cows can have trouble conceiving a new calf the next year. For these reasons, the Yegerlehners are considering other options. These options include selling more bull calves than usual and letting the kept ones nurse from their own mothers. He may also hire someone to raise his calves. This contractor would divert milk from his herd to feed the calves.

Weaning

Calves can be weaned as early as 2 or 3 months old or left to nurse to as long as 9 months old. Ideally, wait to wean the calves until they are at least 4 months old to give their rumens time to develop. If you wean before it is developed, you will have to give the calves a protein source. Most often, calves are weaned at about 7 months old.

Weaning can be a stressful time for both mothers and calves. At weaning, calves are separated from their mothers so they can no longer nurse. Traditionally, calves have been totally separated from their mothers or nurse cows into areas where they cannot see each other. The calves get stressed and spend their time bawling for their mothers instead of eating. The mothers will also bawl for their calves. Mothers will slack on eating for a while, but they will eat when they are hungry.

One way to reduce stress is fenceline weaning. In **fenceline weaning**, calves are separated by a fence into an adjoining pasture where they can still see, hear, and smell their mothers. For animals accustomed to an electric fence, three strands of wire should be enough to keep them separated. The separation is accomplished by moving the mothers into the adjacent pasture and leaving the calves where they are. Be careful when separating the calves and mothers. Keep in mind the principles of blind spots and point of balance. *This was discussed in Chapter 5.*

If you plan to sell calves soon after weaning, make sure you wean them a few weeks before the sale date. Ensure the calves have access to water. Some farmers let their water troughs overflow for the first couple of days to attract the calves' attention and ensure they know where the water is. Check in on the calves to make sure they are eating and are not depressed.

Dairies that use nurse cows can keep their calves nursing for the same amount of time as beef producers, roughly six months or so. If you are still milking your other dairy cows, milk your nurse cows in the barn after weaning.

Veal

Veal was developed as a way to make use of the many dairy bull calves that will not be used for breeding. These calves were considered almost useless in a dairy system, so farmers began slaughtering these calves at a few months old and offering small, lean, tender cuts of meat.

You may have heard the horror stories about conventionally raised veal, how the male calves are taken from their mothers shortly after birth and stuck in dark crates so small they cannot even turn around. These calves are also fed a milk substitute that keeps them anemic and keeps their meat white. Like many conventionally raised cattle, they are given growth hormones to speed growth and antibiotics to fight the infections likely in this type of environment. The backlash against these practices was so strong many Americans refused to eat veal. In recent years, even many of the largest veal producers have moved away from crates and have begun raising small groups of veal calves in pens, a practice known as group housing. In 2007, the American Veal Association passed a resolution calling for all veal calves to be raised in group housing by the end of 2017.

Many grass-fed dairy cattle farmers go further than that to humanely raise veal and instead raise pasture-raised veal. Pasture-raised veal calves are left on their mothers, or on nurse cows, and allowed to live naturally until time for harvest at 5 to 6 months old. To soothe people who are still saddened that even pasture-raised veal calves have to die so young, grass-fed cattle farmers point out these calves are slaughtered at the same age as lambs. You can tell how veal was raised just by looking at it. Pasture-raised veal will be pink or rose-colored, which is why pasture-raised veal is sometimes called rose veal.

Drying off dairy cows

Stop milking dairy cows the last two to four months before calving. Milk production slows greatly at this time as dairy cattle shift their energy intake toward their developing fetuses. If you have a seasonal dairy, meaning all your cows give birth at the same time, you will not milk any cows for at least two months a year. Dairies that use staggered calving seasons can keep milking some animals while drying off those near calving. After the last milking, conventional dairy cows will infuse all four teats with antibiotics to heal any cases of mastitis. Use organic teat sealant solutions to avoid antibiotics, which achieve the same goals as conventional sealants. Drying off is a stressful time for cattle, so monitor them to ensure they are all adapting to the change. Cows' immune systems are weak a few days after drying off and two or three weeks before and after calving. Mineral or other nutritional supplements may be necessary to boost the health of the stressed animals. Cows that get sick during pregnancy can threaten the lives of the fetuses. Keep dry cows in good physical condition — not too fat but not too thin. After drying off, the teats form a keratin plug that prevents new infection. The plug disappears a couple of weeks before calving.

Post Weaning

Newly weaned calves are still growing after weaning, so make sure they still gain weight in this phase. A good forage base will take care of this, though forage's winter or summer dormant periods can slow down weight gain. If you fear your cattle are going to lose weight, use hay or other supplements to keep them gaining weight. The more they can gain in this phase, the easier it will be to finish them when grass growth picks up again.

Throughout the stocker phase, a good average rate of daily gain is 1.5 pounds. You may be able to surpass this during times of peak grass growth. Daily gains will slow when grass goes dormant, but farmers have reported good-quality beef even if their animals gained as little as ½ pound a day during December and January or July and August. For optimum beef quality, researchers of a joint project called "Economic Pasture Based Beef Systems for Appalachia" from the USDA-ARS Appalachian Farming Systems Research Center, in Beaver, West Virginia; Clemson University; Virginia Polytechnic Institute & State University; and West Virginia University, recommended about 1 pound per day during the winter months. Aim for about 2 pounds per day in the finishing phase. Many of the farmers interviewed for this book said they are not able to weigh their animals to monitor precise average daily gains. But as you develop an eye for animals, you will be able to tell if they are losing, gaining, or maintaining weight. The farmers said they try to feed their animals well enough to keep them gaining in winter and to avoid any periods of time where the animals are losing weight.

Grass Finishing: Helping Animals Mature Naturally

Unlike most beef produced in the United States, grass-fed beef is finished mostly on pasture or, in some cases, on pasture with a small percentage of supplements. The biggest challenge for grass finishing is consistency. Grain-fed cattle are finished the same way every time, in the same penned conditions with the same grain diets. Grass-finished animals face more variable environments, from weather to the types of grasses growing, that in many parts of the country can vary widely from day to day.

The best time to target for finishing animals depends on the grasses in your pasture and the time of year when the grass grows fastest. In most places, you can finish your animals on lush spring pastures. In other places, such as

Nebraska or other upper Midwest states, summer grasses are best. Pastures with a high mix of legumes are good finishing pastures. Although alfalfa does not handle grazing particularly well, alfalfa-based pastures can produce gains similar to grain, about 3 pounds per day. Farming experts in your state can help you determine the best time to finish.

The key to finishing high-quality beef animals is to see they are gaining a high average amount of daily weight in the finishing phase. On its website, the AGA defines the finishing phase as the last 200 pounds before slaughter. They should be gaining as much as possible in the finishing phase, at least about 2 pounds per day. These high gains allow the animals to develop intramuscular fat that affects tenderness, juiciness, and flavor.

Many grass-fed cattle farmers try to finish their animals completely on pasture, but the AGA also allows some supplements. In determining which supplements to allow and which to outlaw, AGA decided supplements should be comparable to the major nutritional content of forages. This means supplements should contain energy, fiber, starch, and protein. Because most grasses and legumes contain 20 to 30 percent starch, AGA judges supplements on the criteria of starch-to-fiber ratio. So supplements with a nonfibrous carbohydrate content of 30 percent or less are allowed. Approved supplements include cottonseed hulls, oat hulls, and corn silage cut before it developed into grain. Banned feedstuffs include urea, a non-protein source of nitrogen, and cereal grains. *A more detailed list of approved and banned supplements is available in the Appendix B.* For the most up-to-date information, visit the AGA website at **www.americangrassfed.org**.

Knowing When an Animal is Ready for Slaughter

Grass-fed farmers who keep weight gains high year-round are able to finish in as little as 16 to 18 months, but many cattle need 20 to 22 months and sometimes more than two years. Determining when an animal is finished can be tricky. Many new grass-finishing producers, even those who have raised beef in conventional systems, often slaughter their grass-fed animals too early, before they are truly finished.

Producers with scales can set an ideal target weight and use a portable scale to occasionally weigh their animals to see how they are progressing. Most grass-fed animals are finished at between 1,000 and 1,200 pounds. If you do not have a scale, you might be able to visually determine when an animal is finished. Smaller-framed animals that gain weight quickly might not need to be very heavy, but experienced farmers become talented at telling which animals are ready for slaughter. Look at areas of the body where fat is especially visible, such as the rump, sides, and back. For example, a fat, fully-finished animal will have a rounded rump around the base of the tail head, flat, smooth sides with no indentations, and a flat back. Also check out the brisket area between the front legs. If these sections are getting fat, the animal is probably getting close to finishing.

Ultrasounds

Larger producers sometimes use ultrasounds to determine when their animals are finished. These scans are taken with an ultrasound scanner to record images that measure fat thickness and ribeye area at the 12th rib. The image also gives an idea of the marbling the cows' meat will have. Conducting ultrasounds takes some of the guesswork out of determining which animals are ready to eat.

Completing the Life Cycle

A healthy beef cow on a grass-fed farm can stay productive at least 12 years and often 15, 18, or even more. Some experts would recommend selling after age 12 because after this, the odds go up that the cow will die on the farm and you might not be able to butcher her for hamburger or sell her to someone else who will turn her into hamburger (the price you receive for selling an older or unproductive animal is called **salvage value**).

Most bulls are used for about four or five years before they are sold. After this, the chances of reduced fertility increase. If you have a bull you really like, you may feel it is worth the risk of decreased fertility to keep him longer than four or five years. This will depend mostly on if he is still impregnating a high percentage of cows. If he still seems viable, then try to keep him another year. If conception rates seem to be dropping, or if it seems as though his libido is decreasing, it may be time to get rid of him. You can also have a vet give him a soundness exam, like the one discussed in Chapter 5, a couple of months before the breeding season to make sure he is still up to the job.

It is normal for farmers to feel sadness when taking their cattle to slaughter. Farmers spend lots of time with their herd, and they often form a bond with their animals, which come to trust and depend on the farmer for food. Many farmers often become affectionate toward animals that have been especially good to them, such as a good bull or a cow that delivered many good calves. So naturally, it can be tough to send them off to slaughter. But farmers can take comfort in knowing this is the life cycle for their animals and their animals are dying for a worthy purpose. They can also take comfort in knowing they gave their cattle a good, healthy life as close to natural as possible.

CASE STUDY: SWISS CONNECTION CHEESE

Alan, Mary, and Kate Yegerlehner
Swiss Connection Cheese
Clay City, Indiana
alan@swissconnectioncheese.com
www.swissconnectioncheese.com

Alan and Mary Yegerlehner traded quantity for quality when they gave up their leased grain fields and switched from a conventional dairy to a grass-fed cattle farm on which they make their own raw-milk cheeses.

At nearly every step of the process, they forgo quantity. They are a seasonal dairy, and their dry season is about four months, nearly two months longer than most dairies' dry seasons. They lose quite a bit of milk quantity by using nurse cows, whose milk feeds their calves instead of going into the bulk tank. And, perhaps the most unusual management choice of all, they only milk their cows once a day, which sacrifices a small portion of milk but making a big dent on time spent on this chore.

Their daughter, Kate, who works full-time on the farm, handles much of the animal management. The herd consists of Dutch Belt, Milking Shorthorn and Devon crosses, some of which are descendents of the Jerseys, Holsteins, and Guernseys milked conventionally on the farm before the switch. Their crossbreeding has resulted in smaller-frame cows less than 1,000 pounds at maturity. They keep their cattle on 200 acres by their house in warm weather. In winter, they use neighbors to help them on a cattle drive to an 80-acre pasture about 3½ miles up the road. Neighbors drive ahead and behind the herd or walk alongside while carrying nylon fence. Cattle that get outside the fence quickly want to be back in with the herd.

They get an average of about 25 pounds of milk per cow per day, about 80 percent of the volume they would have gotten if they milked twice daily. They only use about two-thirds of their 75 cows for milk; the others are used as nurse cows.

They have found creative ways to make more money from their byproducts. Some of their bull calves are sold for naturally nursed veal; some of the other males are grown into dairy beef. They buy feeder pigs in the spring that they fatten up in the pasture with whey and skimmed milk. In the fall, the pigs are harvested for pork chops, ground pork, and sausage.

The heart of the business is cheese. They have an on-farm store from which they sell their products. Their products include natural rind cheeses, which means the outer layer of cheesecloth, wax, or hardened rind allows the cheese to "breathe" and develop more complex flavors. They price cheeses by the pound, selling ½-pound wedges and 5-pound or 10-pound wedges, with prices ranging from $10.50 per pound for Colby or aged jack to $16 per pound for fresh mozzarella. In Indiana, as in many states, it is illegal to sell raw milk for human consumption. It is also legal to sell raw-milk cheeses aged for 60 days.

The Yegerlehners began processing their milk on-farm in 2000. They spent about $90,000 remodeling part of their dairy barn, turning it into an on-farm store and a cheese-processing facility with equipment bought from a retiring cheese maker. They use a cheese vat, a pasteurizer on a small number of products, and a vacuum sealer, and they store products in refrigerator trucks and walk-in freezers. They also remodeled their herringbone-style parlor and turned it into a swing parlor with eight milking units, which doubled the number of cows they could milk at a time.

At first, they made 40-pound blocks of pasteurized cheese sealed in plastic. In 2004, Kate attended the famed Terra Madre sustainable food conference in Italy; Alan and Mary attended in 2006. Tasting the cheeses there inspired them to try new methods of making raw-milk cheeses.

"The flavor was out of this world," Alan Yegerlehner said. "Switching from anaerobic plastic to aerobic just made a tremendous difference in flavors. We were making good cheeses in plastic, but to be able to step up to a little more authentic, artisanal touch allowed a lot of those flavors in our milk to be expressed even more."

• Chapter 9 •

Beef and Dairy Processing

TERMS TO KNOW

Live weight: How much an animal weighs just before it is slaughtered. Another term for live weight is "on the hoof."

Hanging weight: The number of pounds beef weighs while hanging on a rail in the cooler. This is after excess fat and bones have been removed. It is usually between 55 and 68 percent of the live weight. Most beef is priced based on the hanging weight. Hanging weight is also sometimes called "dress weight" or "dress percentage."

Trim: Meat leftover after the butcher removes the best cuts. Trim is used for ground beef, roasts, or other lower-cost cuts.

Pasteurization: The process of heating milk to kill bacteria

Raw milk: Milk that has not been pasteurized

Aging: Letting beef hang in the freezer for several days so natural enzymes can break down connective tissue that makes meat tough.

The end goal of grass-fed beef and dairy cattle farming is to harvest meat and milk sold as products customers want. *In this chapter, you will learn about processing.* In beef production, "processing" usually refers to harvesting (a term often used instead of slaughtering) the animals and cutting and packaging the meat. For dairy production, milk processing involves pasteurization or turning milk into other packaged products. In some states, it is also legal to sell bottled raw fluid milk. Because grass-fed products are a niche market, many farmers have found that they can get the most value from their products by making processing arrangements that allow them to sell their meat or milk directly to their customers. This is a better option than selling their animals to feedlots or selling their milk on the commodities market where quantity can be more important than quality.

Grass-fed dairy farmers must find a way to get the extra value from their pasture-raised animals. Selling on the commodities market will not provide the financial rewards you can get from more discerning customers who are looking for a higher-quality product. Dairies must decide what kinds of products they want to sell. Do they simply want to bottle it and sell it as raw or as pasteurized grass-fed milk, or do they want to convert the milk into cheeses, yogurts, or other products that may also have added value to customers? Either way, they must either find the best processor to fit their goals or learn how to do these tasks themselves. A growing number of small dairies are processing their milk themselves, either by bottling on the farm or by making their own products such as cheeses and yogurts. Dairy producers who do not process their milk on their farms must pay someone to process and package their products for them, participate in a cooperative that has a processing arrangement, sell to a processing company that buys milk to make its own products. *These topics will be explored later in the chapter.*

Some farmers butcher their animals themselves, but most must take their animals off-farm to be killed and butchered. Usually, they load the animals into a cattle trailer and ship them to a slaughterhouse. When the animals arrive at these processing facilities, they are usually stunned unconscious, then hung up and cut so they bleed to death. Then, they are skinned and their heads and organs are removed, and the remaining meat is split in half (these halves are called sides). These sides can be cut further into familiar retail cuts.

Knowing the Laws

Farmers must follow many laws at the federal, state, and local level. Because lawmakers are concerned about the safety of the food people eat, this is especially true in regards to processing. There are laws about how to clean equipment in slaughterhouses and dairy barns and how to store meat in butchers' freezers. These laws are intended to keep people safe, and many of the guidelines are necessary and effective. But many laws are also aimed at large-scale productions, and the fine print or strict requirements can cause headaches for smaller farmers and processors. It is important to know your state's laws about selling your products. The best way to find out what you can and cannot do is to ask someone who works at your state Department of Agriculture or a local extension office.

Beef

Where beef is slaughtered and butchered and where and how dairy products are processed affect where and to whom you can sell. Beef slaughtering facilities must be inspected by government officials. If you want to sell individual cuts to people, restaurants, or retail outlets, the carcass must be inspected. State and federally inspected processing plants have the same requirements to ensure meat is safe to eat. To sell your meat across state lines, you must slaughter your animals at a federally inspected plant, which means, among other things, a federal inspector will check each animal. Individual states may have other laws regarding the sale of meat processed in state-inspected plants. For example, in Kentucky, you can only sell individual cuts of beef if the carcass is federally inspected, but in Indiana, you can sell individual cuts if it is state or federally inspected. If you have questions, the people you work with along the way, at the slaughterhouse, butcher, and retail outlets, will have a good idea of the laws in your state. Government officials and extension agents can also be reached for help.

Another option you have is to use facilities called **custom-exempt plants**. At these plants, only the facilities (not the cattle) are inspected by state inspectors. Often these smaller, low-volume plants cannot justify the cost of the plant upgrades required under federal regulations. The plants are only allowed to slaughter and process animals for the use of the owner and non-paying guests. These facilities are inspected by federal and state inspectors, but they cannot afford to keep around regular inspectors the way larger plants do. If you are raising cattle to supply beef only for your family, or if you reach a deal with customers to buy a percentage (usually one quarter or one half) of one of your animals while it is alive, you can slaughter the animal at a custom-exempt facility. Usually, the cattle farmer keeps the animal until it is time to transport it to the slaughtering facility. The percentage of the animal owned is sometimes referred to as a share.

Theoretically, a large number of families can own shares in a beef animal, but some states limit the number of people who can own shares in a single animal on the grounds that if more than a small number of people own shares, you are just skirting the laws about selling individual cuts from an animal inspected at a facility with an on-site inspector.

Dairies

You may need permits or licenses to sell milk off your farm or to process milk on your farm. Check with your state Department of Agriculture to see which rules apply to you.

One of the most hotly debated legal issues for grass-fed dairies is raw milk sales. Although some dairies prefer to sell pasteurized milk and dairy products, others are passionate about raw, unpasteurized milk. Most milk sold in the United States is pasteurized, or heated up enough to kill bacteria. This practice was started decades ago when unsanitary conditions on farms encouraged the spread of diseases such as brucellosis. But supporters of raw milk sales, including dairy farmers interviewed for this book, say sanitary conditions on modern-day family farms have improved greatly, which makes pasteurization unnecessary. They say pasteurization kills beneficial bacteria, enzymes, and minerals. One organization that advocates for raw milk is the Weston A. Price Foundation (**www.westonaprice.org**). Many health experts disagree and say drinking raw milk is risky because of the chance milk could carry E. coli bacteria or salmonella.

Because of these health concerns, many states do not allow the sale of raw milk. It is legal to drink raw milk from your own animals, so in some states, farmers get around these restrictions by selling milk shares. Owning a milk share means you actually own a percentage of the animal. Some farmers who sell shares also charge a housing fee to shareowners. Shareowners can then come to the farm at pre-arranged times to pick up milk from their cow. However, some states specifically outlaw these shares. Again, it is very important to check with your state Department of Agriculture to determine which rules you must follow.

To get an idea of the national picture involving raw milk sales, the Farm-to-Consumer Legal Defense Fund, a nonprofit group that pools resources to fight for the rights of family farms and consumers, has an interactive map that provides an overview of raw milk laws around the country on its website (**www.farmtoconsumer.org**). It is legal to sell raw milk in Pennsylvania, Connecticut, Maine, New Hampshire, Idaho, Washington, California, Arizona, South Carolina, and New Mexico. It is illegal in Montana, Nevada, Louisiana, Iowa, Wisconsin, West Virginia, Maryland, Delaware, Rhode Island, Washington, D.C., and New Jersey. This interactive map can give you a look at the big picture regarding raw milk sales in the United States, but — this cannot be stressed enough — contact your state department of agriculture to be sure you know what laws you must follow.

How to Choose a Slaughter House

Another big issue for beef farmers is finding a place to have their animals slaughtered. There are not many facilities, and many of them cater to large producers. The challenge for beef farmers is finding a facility within a reasonable distance from their farm that can also do all the necessary work. Being loaded into cattle trailers and hauled many miles down the road is stressful to animals. The shorter the trip, the better, but some farmers have no choice but to truck their animals hundreds of miles away. Once you find a facility to harvest your animals, you usually need an appointment because even small- and medium-size facilities often have waiting lists. These appointments may need to be made many weeks in advance.

Some slaughterhouses also have a meat cutter on site, but some only kill. If you find a facility that only kills, your cattle may have to be taken by refrigerated truck to a place where they can be cut and packaged. Find these facilities through your state Department of Agriculture, which keeps a list of processors, including whether they only kill, only cut, or both.

Be sure to visit slaughterhouses when you are choosing the facility you want to sell your cattle to. When you visit, call ahead and let them know you are coming because they may be more accommodating if they were expecting you than if you drop in unannounced. When you visit, talk to people and see what kind of feeling you get from the employees. You want to work with people who respect the animals and take pride in the job they do. Slaughterhouses are noisy places, but watch the way people work. They should seem calm and in control. Look for facilities that adhere to the same principles as you did when designing your handling facilities, for example, rounded walkways aimed at working with the animals' natural instincts.

Organic slaughter and butchering

Producers who want the organic label must be careful in choosing processors. Organic-labeled products must be processed and handled by organic-certified operators — that means slaughterers, milk processors, and packers. These facilities often produce conventionally raised beef as well, so they must take extra precautions to ensure organically raised beef is kept separate throughout the process. One of the key attractions to organic food is it is pure of the chemicals found in conventional agriculture, and this must continue through slaughter and processing. Such precautions to ensure this purity during the slaughtering stage include handling the organic animals first before the facilities have been exposed to conventionally raised animals that day and keeping the different types of animals in separate holding areas.

Mobile processors

Mobile processors have been growing in popularity because they offer an alternative to the long wait times at larger facilities and because they eliminate the stress of transport on live animals. These are often mobile trailers where animals can be killed, cut into sides and quarters, and chilled. Sometimes, the carcasses have to be taken to a butcher off-premise for final cutting and wrapping, and these trucks can do this for you. Some are USDA-inspected, and some are organic. Local farming experts will know if a mobile processor could be available to you.

Meat Cutting

A meat cutter, also called a butcher, cuts your beef into retail cuts. This section provides an overview of how this is accomplished. Here are some beef processing terms you should know:

- **Hanging weight:** the weight of a side of beef as it hangs on the rail in a meat cooler. This is after excess fat and bones have been removed. Most beef is priced based on the hanging weight. Hanging weight is also sometimes called "dress weight" or "dress percentage."
- **Carcass yield**: Slaughtered animals will yield between 55 and 68 percent of their live weight for meat (the rest of the animal was blood, guts, bones, hide, and other material). The exact percentage will vary depending on breed, sex, and body condition of the animal.
- **Trim**: Meat left over after the butcher removes the best cuts. Trim is used for ground beef, roasts, or other lower-cost cuts.

Aging

When choosing a butcher, make sure he or she has room to hang your beef for as long as you need. Meat needs to be aged after slaughter and before it is cut. **Aging** means letting the sides or quarters hang in the freezer for several days so natural enzymes can break down connective tissues that make meat tough.

There are two ways to do this:

- Wet aged meat is stored in vacuum packaging and allowed to age in its own juices, usually for about seven days. Most mass-produced beef is wet aged. It is a cheaper and quicker process than dry aging. Unlike in the dry aging process, you do not lose meat volume to shrinkage. There are some niche producers who also prefer wet aging because it preserves many of the meat's juices.
- Dry aging is the preferred method of most small farmers because those who do it consider this process to produce a more intense flavor. Dry aged beef is hung in the open air. Tough muscle tissues are allowed to break down, which produces tenderer meat. Dry aged beef usually hangs for ten to 14 days. Fat on the carcass determines how long it hangs. If there is not much fat, it cannot hang as long. Beef shrinks about 15 percent during the dry aging process.

Many grass-fed beef farmers complain about laws they must follow to cool their meat. Laws require meat to be hung in the freezer within an hour of slaughter. But for a couple of hours after cattle are slaughtered, their muscle tissues release an enzyme that keeps their tissues from tightening up, which essentially tenderizes the meat. Meat that cools down soon after slaughter does not reap the benefits of this enzyme. This is not as big an issue for grain-fed meat because these carcasses contain more fat, which insulates the meat and prevents it from cooling so fast the enzyme does not work. Grass-finished beef does not have this protective layer of fat, so the enzyme is rendered ineffective and the meat becomes tough.

USDA grades

You can pay the USDA to assign each animal's beef a grade that indicates the tenderness and flavor of the meat. The grading system is based on age (younger beef is more tender) and marbling (fat that grows between muscle fibers, a factor in flavor and juiciness). Under this system, the greater amount of marbling and the younger the beef, the better the grade. Grass-fed beef shows less marbling than grain-fed beef, even between two products comparable in tenderness. The grading system usually favors grain-finishing, so many grass-fed producers choose not to get their meat graded, and many producers explain this decision on their websites.

Because some grass-fed producers do choose to get their beef graded, here is a look at the USDA grades. The three main grades of beef are prime (considered the best grade), choice, and select. There are also degrees of grades: high, medium, and low. Medium grades are rarely included, and high and low grades are given most often.

- **Prime**: considered the best-quality meat. It has the most marbling and comes from animals finished at younger ages. Prime cuts are the rarest cuts and are usually only found at top-quality restaurants. These cuts are also the most expensive for consumers to buy.
- **Choice**: usually considered a notch below prime and has less marbling. Choice cuts are considered very good quality but more affordable than prime cuts.
- **Select**: this cut is leaner than choice or prime, which is common for grass-finished animals. Even though grass-fed beef often grades lower than grain-finished meat, the products can be comparable in quality.

There are other grades: standard, commercial, utility, cutter, and canner, but meat of these levels of quality are not sold retail. Lower grades are used in low-quality products such as canned meats.

Know your cuts of meat

Customers can buy the whole animal, sides (half the animal cut lengthwise), or quarters. You can also sell your animal **on the hoof**, which means you sell the animal or shares of the animal when it is still alive. Beef from animals sold on the hoof is sold in bulk, usually in halves or quarters. You can also sell quarters of the animal, such as a forequarter or a hindquarter. Selling in quarters can be difficult because the best cuts come from the middle and back of the animal, and it can be hard to get rid of cuts that come only from the front of the animal. Selling split sides is a way to sell a quarter of the animal and mix up the cuts. A split side is half the meat from a side of beef including cuts from the front and back.

Not every beef animal is cut the same way. You need to be able to tell your meat cutter some basic things about the cuts you want. If you are unsure, the cutter will be able to offer advice and suggestions. As you get used to the process, and as you get feedback from your customers, you will start to get a feel for what customers want and what they do not. Individual customers may also want different things. Some want a small number of thick steak cuts; others may want more steaks with thinner cuts. If you sell a whole side of beef, it can be cut exactly the way the customer wants it. If you sell split sides, it is difficult to do custom cuts unless both customers want the same thing.

A beef carcass is automatically split into sides, but you can also think of it as being divided into quarters, two forequarters in the front and two hindquarters in the back. In general, the best, most tender steak cuts come from the rear quarter of the animal, except for the ribeye, which comes from the front part of the middle section and is included with the forequarter. The less-tender cuts come from the heavily exercised areas, including the shoulders and rump areas.

Courtesy of the California Department of Food and Agriculture.

From the chuck area at the top of the forequarters, you get several roasts (chuck eye, seven-bone, boneless chuck pot roast, cross rib pot roast, under blade pot roast, blade roast), top blade steak, flat iron steak, and short rib steak. This is where a lot of ground beef and stew beef come from.

From the ribs, you get rib roasts, rib steak, rib eye roast, rib eye roast, and back ribs.

From the lower forequarter, you get briskets (whole or half) or a shank cross cut. This is also where ground-beef-type cuts come from.

From the lower midsection of the underside, you get flank steak, flank steak rolls, or fajita meat.

From the top middle (short loin and sirloin), you get most of the expensive cuts: T-bone steak, sirloin steak, porterhouse steak, filet mignon (tenderloin steak and tenderloin roast), top loin steak.

From the rump/round area: rump roast, top round steak, tip steak, tip roast, kabob cubes, eye round steak, bottom round roast

Wrapping

Also, consider how butchers wrap their products. Some wrap them in plastic and cover them with white paper. If you are selling individual cuts, you may want a butcher who uses transparent vacuum wrapping to allow customers to see the meat inside the package before they buy.

Pricing

Base your price on the desired profit you want to make and take into account your expenses per pound. Put a monetary value on your time and labor as well. You can set your prices by the pound, either by live weight, hanging weight, or dress percentage, or decide on a number you think is fair, if customers will pay it.

You can determine a range to charge for your meat by checking prices farmers in your region with similar operations will charge. This will depend in part on what customers in your area will pay. For example, you will be more likely to find high-paying customers in New York City or Los Angeles than in Metropolis, Illinois. Like with any business, you will need to sell for more than what you pay in expenses. So figure what you spent during the year on seeds, fertilizers, vaccines, gasoline for farm equipment, and anything else you needed to keep your farm going, and use that to determine how much you need to make from each animal to be profitable.

People who sell shares or in bulk usually price by the pound according to live weight, but you can also sell by dress weight because you probably will get differing amounts of meat from two live animals that weigh the same. People who sell individual cuts often sell by the pound, but each pound is not created equally. You need to sell your steaks high enough per pound to balance out the lower price per pound of hamburger.

Dairy Processing

When you produce milk, you can drink it yourself, sell it, or turn it into other products to eat or sell. Milk must be cooled quickly, or it will spoil. Raw milk has a shelf life of a week or a little more. Milk can be processed into other products to help preserve it. Farmers selling milk conventionally often contract to market their milk through a milk-marketing cooperative. More than 85 percent of milk in the United States is marketed through milk-marketing cooperatives, which may bottle or process the milk themselves under a farmer-owned brand (one you may know is "Land-O-Lakes" butter), or sell the bulk milk to private processing companies that will package it as fluid milk or other products and sell it around the country. These farmers pump their milk into a bulk tank. Contracted milk haulers collect from these bulk tanks at least every other day.

Conventionally sold milk is always pasteurized. Pasteurization is the process of heating milk to kill bacteria. There are three main methods of pasteurization:

- **Batch pasteurization:** This old-school method of pasteurization involves heating milk in a vat to 145 degrees and maintaining the heat for 30 minutes. Batch pasteurization is also used on milk used for cheese, yogurt, and other value-added products.

- **High Temperature/Short Time (HTST) pasteurization**: Milk is heated to least 161 degrees for at least 15 seconds. This is the most common method used in the United States.

- **Ultrapasteurization (UHT)**: Milk is heated to 280 degrees for two seconds.

  If milk is kept unopened after UHT, it does not have to be refrigerated for six months or more and can be shipped warm.

Many on-farm bottling dairies pasteurize their milk because in many states it is required by law. These dairies can pasteurize their products on the farm using a pasteurizing machine. The HTST method is usually preferred because it is quicker than batch pasteurizing but does not kill all bacteria, harmful or beneficial, the way UHT does. Buy pasteurizers from specialty dairy suppliers, as small as 2 gallons or as big as 300 gallons. You could also pasteurize milk on your stove if you cook it at the right time and temperature.

Conventionally sold milk is **homogenized**, which means a machine is used to break up fat globules into small pieces so the cream will not rise to the top. Many grass-fed dairy farmers choose not to homogenize their dairy products because they prefer the more natural product with cream at the top. However, some grass-fed creameries do offer customers the option of getting homogenized grass-fed milk.

Whole milk is milk without the fat removed. Whole milk has at least 3.25 percent butterfat. The milk from Jerseys can contain 5 percent milkfat or more. This fat can also be used to make other products such as butter or cheese. Not all grass-fed dairies sell raw milk or just whole milk. Some farmers make grass-fed skim milk or reduced fat milk such as 2-percent milk, 1-percent milk, and fat-free milk. Some grass-fed dairies also sell chocolate milk.

You can make many other products from milk as well. Making these products requires extra equipment and considerable skill, so take time to study each product and to learn how to make them from someone offering classes or farm internships. ATTRA has a list of farm internships at **http://attra.ncat.org/attra-pub/internships**. Although the number of dairies is small even in the niche world of grass-based farming, enough grass-fed products are available now, so your customers will demand the best quality. The extra time and financial investment could be worth it. Discerning customers will pay more for farmstead dairy products.

The following are some of the dairy items you can create using milk from your grass-fed cattle:

- **Cheese**: this is usually made from whole milk, but you can also make it from skim milk or cream. Cheese is made by using enzymes or cultures to separate the solids, called curds, from liquids, called whey. The solids are pressed into a cheese shape and, depending on the type of cheese you hope to create, some are allowed to age for many months or even years. Cheeses made from unpasteurized milk must age for 60 days. Cheese can be cut into smaller pieces for sale or sold in big wheels.

- **Whey**: a byproduct of cheese making. It can be sold as a health product or used to feed hogs, which can also be sold later for meat.

- **Cream**: a dairy product made from butterfat. Cream from grass-fed cattle usually has a more yellow color than cream from a grain-fed cow. This is a reflection of the extra beta carotene from the all-forage diet.

- **Butter**: a product made from milkfat. Under federal law, it must be at least 80 percent milkfat. Some dairies sell unpasteurized butter. Butter can be frozen and sold year-round.

- **Whipped butter**: the same as regular butter, but it is whipped and easier to spread

- **Yogurt**: fermented with live bacteria cultures. These cultures make them easier to digest and boost the immune system.

- **Ice cream**: ice cream made from grass-fed milk, especially without artificial colors and flavors, is a healthier option than mass-produced ice cream.

- **Sour cream**: made by adding bacteria that sour the cream

Selling to other companies

You will not have to worry about processing if you sell to other companies because these companies will process your milk into whatever products they sell. A handful of small creameries buy grass-fed milk. Local agriculture experts such as extension agents should be able to help you find them. You may also find a large company with a niche grass-fed line. One example of this is Grass Point Farms, a line from the Organic Farm Marketing company. Selling to other companies usually involves a contract, and you must produce your milk to their standards. Often, you are not allowed to use your milk for any other sales purpose, and you must give it all to the company you are contracting with.

Cooperatives

You may also be able to join a milk-marketing cooperative specializing in grass-fed milk. **Cooperatives** are groups of farmers who share resources and market their collective products under one brand name. An existing dairy cooperative probably already has an arrangement with a processing company. An example of a cooperative is PastureLand, a group of grass-fed organic farmers based in Minnesota. Beef farmers can also form cooperatives. An example is the brand Wisconsin Meadows, a group of Wisconsin farmers.

Farmstead processing

As mentioned earlier, many farmers who target niche markets find they can get the best price for their products by producing it and marketing it themselves. If you want to make a living selling dairy products, on-farm processing is a big undertaking that will require an additional investment in equipment and training. Some dairies also employ an additional person to do the actual processing. Weigh the costs of these undertakings against the goals of your farm. It easily could cost a few thousand dollars to buy equipment such as a cheese vat, a cream separator, and packaging equipment. If you just want to make products yourself, there are companies that specialize in less costly, kitchen-size equipment, such as New England Cheesemaking Supply Company (**www.cheesemaking.com**) and Glengarry Cheesemaking and Dairy Supply (**www.glengarrycheesemaking.on.ca**). You may be able to find deals on used dairy equipment at auctions or online. You will also need packaging equipment and coolers or freezers to store your products.

If you process on your farm, it would be a good idea to take classes to learn these skills. Farmers or extension offices in your area may know of classes offered or they may offer on-farm apprenticeships. Many dairy farms that process on-site have a full-time employee to run the processing side of the business. You might even find this to be a good income for another family member. It takes time to learn these processes, so be comfortable in your livestock management skills before you take on this second endeavor.

Co-packing

If you wanted to market your own line of dairy products, but you do not have the know-how or the desire to make your own, pay a local cheese plant to do it for you. A company that packages products to your specifications is called a **co-packer**. Send these companies your milk and they turn it into cheese (or whatever dairy product you choose to make) and put your label on it. You would then be responsible for marketing the product yourself.

You will learn more about marketing strategies in the next chapter.

CASE STUDY: WHITE OAK PASTURES

Will Harris, owner
White Oak Pastures
Bluffton, Georgia
http://whiteoakpastures.com

Will Harris used to take his beef cattle to a slaughtering facility 9 miles from his farm in Bluffton, Georgia. When the facility closed, his next-best option was 90 miles away. Although many beef producers truck their animals to slaughterhouses at even greater distances, Harris decided this plant was just too far away from his farm. He knows not everyone will have a choice in this because of the limited availability of USDA-inspected facilities, but he did not like putting his animals through the stressful trek to a slaughterhouse 100 miles away. He said there is no set mileage limit for hauling animals, but "the closer the better." So Harris, whose grass-fed meats can be found in several states in the eastern United States, did what only a handful of producers have done — he built his own processing plant.

The 5,300-square-foot plant, which cost $2.2 million dollars, opened in 2008. It is USDA-inspected, which allows him to sell individual cuts and products in other states. The plant employs about 25 people and harvests about 18 animals per day, including many from farms in the surrounding area. Harris built it with sustainable features so nothing is wasted. Solar panels provide energy to heat water, which is then recycled for irrigation; waste products such as bones and internal organs are used as fertilizer; hides are sold for leather.

His processing facility adheres to principles Harris himself searched for when choosing facilities in the past. His farm is certified by the Certified Humane Raised and Handled program, which means his animals were given the space and environment to engage in natural behaviors from birth to slaughter and were not given antibiotics or hormones. His facilities are designed to keep the animals calm.

"The animals are never excited in the process of being dispatched," Harris said. "First of all, it is the right thing to do. We human beings assume dominion over these animals, and that makes us responsible for the stewardship of the animals. Furthermore, when you panic an animal and it goes into that fight or flight mode, it releases adrenaline into the system, and it literally changes the pH in the muscle tissue. And that changes the eating quality of the beef.

"We do it one at a time, the old-fashioned way: a man with a knife. It is just a professional man with a knife generating a quality product for which a sophisticated customer will pay you a premium for. We are professional cowboys, and we know how to work cattle. It is a full-time job for everybody who works in my plant. It is what they do for a living.

"Every farmer who does not own their own plant, it is incumbent upon them to go and be there to see that the processor does it correctly."

Harris runs about 650 cattle on 2,000 acres, which gives him the largest certified organic farm in Georgia. He also has a handful of associate producers raising grass-fed cattle for him. He weans calves at about 8 months old and fattens to between 1,000 and 1,100 pounds when they are about 22 months old.

He takes advantage of the South's warm climate to provide beef year-round with a calving season that lasts from September to April. His grazing system depends on warm-season perennials including Bermuda grass, Johnson grass, and bahia grass. In winter, he overseeds cool-season annuals, including rye grass and clover. He also uses a sheep herd to graze down undesirable plants.

He sells much of his meat to the grocery chains Whole Foods and Publix and to upscale Atlanta restaurants, which charge a premium for grass-fed beef. Having his own processing facility gives him even more control over the final product, which ultimately depends on quality. If you are targeting your product to an upscale market, the final product must be as good as it can be.

• Chapter 10 •
Marketing

TERMS TO KNOW:

Community Supported Agriculture: A system in which community members pay for "shares" of the food a farmer is expected to produce in the upcoming year

Beef pooling: also called "cow pooling," this is another shareholder system in which one customer is in charge of coordinating orders for a group of people who buy in bulk. These pools ensure producers have an outlet for selling their meat.

Milk shares: also called herd shares, is an arrangement that allows people to buy raw milk. In some states, these shares are explicitly illegal.

Direct marketing: Selling meat or milk products directly to the customers who consume them

Today's beef and dairy industries are built on the promise of cheap food. Most customers do not know where their food came from or how it was produced. To make such a large amount of food available at low prices, the industry came to be dominated by large producers that could make meat and milk quickly in part by giving animals relatively small individual spaces. This is how most beef came to be produced on feedlots where hundreds of animals are fattened quickly on grain and hormones and given antibiotics to fight off the likely infections from the crowded conditions and unnatural diets.

But a growing number of consumers are concerned with the way their food is produced. They want healthy products not laced with extra hormones or antibiotics that might be unnecessary, and they want to know the animals that provided their food were treated kindly. Many of these customers are also willing to pay more for such assurances.

Many grass-fed farmers market their products themselves. Marketing involves capturing the attention of customers who value what your products offer. The name you give your farm, the way you describe your products, and where you sell products all fall under the category of marketing. If you decide to expand your farm and sell your products, there are a few key selling points for grass-fed beef and dairy products:

- **The nutritional content**: Some customers will be attracted to the additional nutritional content from raising cattle on pasture-based diets. The CLAs, the omega-3s, and vitamins are all selling points, as are the lack of unnecessary antibiotics and hormones. Grass-fed milk and butter will be different colors than the mass-produced stuff (though the mass-marketed products often add colors to mimic the natural look). Conventionally produced milk and meat often loses nutritional content, which is why proponents of raw milk call it "real milk."

- **Taste**: Grass-fed beef and dairy products taste different than the mass-produced alternatives. They even offer unique flavors resulting from the specific soil and vegetation on each farm. Some people will like the taste of grain-fed products better, but many others prefer the grass-fed flavor. Few people who buy or sell grass-fed beef know what meat tasted like before grain finishing became the norm in the middle of the last century, but savvy marketers still point out their grass-fed products taste like beef used to.

- **Happy cows**: After all the attention drawn to the living conditions on industrial feedlots, some people will pay more for the assurance that the products they are buying came from cattle allowed to live more humane, natural lives than those provided on factory farms. Even veal, an often controversial product when raised conventionally, is marketable when people know the calves were left with their mothers out in the fields. The way your cattle are allowed to roam outside and the green of your pastures are selling points for your products. Even describing the grasses on your farm can help your customers paint a picture in their heads of where your products come from.
- **You, the farmer**: Many customers will pay more to support smaller farmers because people like to know they are sustaining families. If a family has farmed for generations, this is often mentioned on product labels. The idea of sensible, sustainable farming practiced for generations resonates with people.
- **Cattle breeds**: Angus cattle are probably the best-known beef breed, and people expect meat from Angus animals to be high quality. But any breed you mention can capture your customers' attention. If you use a lesser-known breed, this will also be interesting to people and may capture their imaginations even more.

Labeling/Packaging

If you sell your product to others, your label must be approved by the government. If you sell across state lines, the label must be approved by federal officials. If you plan to sell your product only in the state you live, you will need to contact your State Department of Agriculture for the labeling requirements and application process. Employees at the facilities where you have your animals slaughtered or butchered will be able to give you an idea of what you will have to do.

The USDA's Food Safety and Inspection Service (FSIS) oversees products sold across state lines. The FSIS aims to ensure products live up to the claims made on their labels. The USDA allows businesses to make marketing claims, for example, grass-fed. These claims are evaluated on a case-by-case basis. If you want to label your products as grass-fed, submit a detailed written protocol on how the animals are raised, a signed affidavit declaring the specifics of the production claim, and an explanation of the animals' diet (including, for example, what the animals are fed during long periods of severe weather). Find more information about the label application process at the FSIS' website (**www.fsis.usda.gov/Regulations_&_Policies/Label_Application_Guidance/index.asp**). Another informative document is "A Guide to Federal Food Labeling Requirements for Meat and Poultry Products," at **www.fsis.usda.gov/PDF/Labeling_Requirements_Guide.pdf**.

If you intentionally misbrand a product, the FSIS can impose penalties including fines, product recalls, or a ban on selling your product. If the FSIS believes your label needs corrections or clarifications, it will provide suggestions.

Another good resource is the California State University–Chico site for grass-fed farmers seeking labeling information. The site (**www.csuchico.edu/agr/grassfedbeef**) includes example labels, a guide to creating your own label, and links to marketing research, as well to the United States Standards for Livestock and Meat Marketing Claims.

There a few terms you can use to describe your products according to USDA standards. Here are some examples:

- **No hormones administered**: this lets consumers know you did not use the growth hormones often given to cattle in industrial operations. (Hormones are not approved for use in poultry, pork, veal calves, or exotic species such as bison and goats, and there additional requirements if you say "no hormones administered" on products from these species.) You cannot say "no hormones" because animals naturally produce their own hormones.
- **No antibiotics administered**: this lets consumers know you did not use the growth hormones, including a type called ionophores, often given to cattle in industrial operations.
- **Natural**: a broad term for "minimally processed and containing no artificial ingredients." Minimally processed means processing does not fundamentally alter the raw product, which means no chemicals, preservatives, or artificial coloring were added to the meat during or after processing; the label must clarify this. The USDA does not perform additional inspections for meat labeled "natural."

Getting Certified

Some customers need the assurance of third-party certifiers before they will pay the premiums for organic or grass-fed products. These certifications are most important for customers who do not buy directly from farmers but at specialty grocery stores or restaurants.

It is not required, but you can have your production claims certified by a third-party agency. AMS' Audit, Review, and Compliance branch (**www.ams.usda.gov/ARCaudits**) has a link to a list of accredited agencies at **www.ams.usda.gov/AMSv1.0/ams.fetchTemplateData.do?template =TemplateD&navID=GradingCertificationandVerfication&leftNav= GradingCertificationandVerfication&page=LSISO65Program**.

Send in your certification papers when you apply for your label application.

The American GrassFed Association offers certification to ensure customers animals were kept outside, raised on pasture-based diets, and were not given growth hormones or antibiotics. To be certified by the AGA, you must be a member. Find membership info, a list of steps required for certification, and a link to the application on the AGA's website at **www.americangrassfed.org**. If you apply for certification, your farm will be audited to determine if you qualify. If you do not qualify, the AGA will send you a list of adjustments so you can meet compliance.

Organic certification

Products labeled as organic must not include any hormones or synthetic chemicals, including fertilizers, pesticides, or medicines such as antibiotics. To use this label, you must follow the rules of the National Organic Program and be certified by an accredited agency. There is a complete list of certifiers on the NOP website (**www.ams.usda.gov/AMSv1.0/NOP**).

To become certified, submit information to a certifying agent including a list of substances applied to your land over the previous three years, the ways you plan to use the land you want certified, and an official organic plan. An **organic plan** lays out how you will run the organic farm, including what types of feed and fertilizers will be used and where these substances come from. The plan must even include details on the record-keeping system you plan to use. The ATTRA website has a sample template you can use for your plan at **http://attra.ncat.org/attra-pub/OSPtemplates.html**. Hang on to records about production, harvesting, and handling your organic products.

When you apply for organic certification, an inspector will visit your farm, observe how you do things, and write a report about this inspection. The certifying agency will review your application and the report to determine if you should be certified. If you are certified, you will be inspected once a year and will have to provide updated information about your organic management practices, such as the substances you have applied to your field. Your certifying agent is also authorized to conduct unannounced inspections.

Farmers who sell less than $5,000 worth of products per year are not required to obtain certification, but they must keep records of their production practices just in case.

Other certifications

There are other agencies you can use on your products whose certification may make them more appealing to customers:

- American Humane Certified (**www.americanhumane.org**) is an independent verifier of humane treatment of farm animals.
- Certified Naturally Grown (**www.naturallygrown.org**) is an organization that adheres to principles similar to those of the National Organic Program but seeks to reduce much of the government paperwork and production costs that discourage many small farmers from seeking certification.

Selling Options

You have two options for selling your products: You can sell directly to the customers who will eat them, or you can sell to restaurants, retail outlets, or distributors who will sell to these customers. Selling directly to customers is called **direct marketing**. The advantages of direct marketing are you can talk to your customers about your product and answers questions for them. Beef farmers who direct market can either sell individual cuts of meat or in bulk.

Bulk sales

If you are starting small and finishing just a handful of animals, start with bulk sales, which allow you to get rid of the whole animal. You can sell:

- The whole animal
- A side of beef (half the animal)
- Quarters (either the half the front or half the back of the animal)
- Split sides or split quarters

You can also sell variety packages that include a few steaks and a few pounds of ground beef and stew meat. If you decide to do this, offer a small enough package that households can afford it and have room in the freezer to store it. An empty freezer in a typical refrigerator can hold about 50 pounds of meat. Obviously, families keep more than meat in their freezer. Some farmers sell bulk packages of 30 pounds or less, but many customers who buy in bulk will need a stand-alone chest or freezer.

Individual cuts

You may be able to find outlets to sell individual cuts. Individual cuts are what you would buy at a grocery, supermarket, or restaurant, one steak for example. Places where you could sell individual cuts include restaurants, which are often eager to buy the best cuts of locally produced beef. You may also be able to sell individual cuts to groceries or to retail outlets that specialize in locally produced farm products. For experienced farmers who have a quality product, a food service business is a good outlet. The downside of selling is it is often hard to get rid of the lower-quality cuts because the most expensive cuts will sell out quickest. Many small farmers find it hard to justify selling individual cuts because customers want the steaks but may not be interested in the stew meat or the tougher cuts. Customers may not appreciate the difference in the quality of your hamburger, which will probably cost more per pound than hamburger at a chain grocery. If you have trouble getting rid of your hamburger, try giving out samples to customers who purchase your more expensive cuts. If you are doing a good job raising your animals, they will notice the difference compared to the cheaper options.

Where to Sell

Most grass-fed cattle farmers will sell their products directly to customers.

A small number have been able to get their products into chain supermarkets.

Bulk sales are most common. Locally owned restaurants and groceries are more likely than chains to buy grass-fed products. It is better to call for an appointment than to show up unannounced.

Farmers markets

Farmers markets are places where groups of farmers gather to sell their products. They are usually held outdoors on weekends in public places. Most vendors at farmers markets sell produce, but beef and dairy farmers also set up sometimes. Farmers markets are usually held from spring through the fall.

Farmers markets are a valuable tool for building a customer base. You can meet potential customers and answer any questions they may have. To entice customers, offer information sheets about the nutritional benefits of your products. You could also hand out free samples to people so they can see for themselves how tasty your products are. Be sure to check with someone at your market to see if you need an extra state license to give out samples.

To sell at a farmers market, you usually need to apply to the person who runs it. You also may have to interview with someone or pay an application fee. Many markets only take applications early in the spring before they open. It might be a good idea to apply to multiple markets because you may not be accepted to one of your choice. Find options on the USDA National Farmers Market Directory (**www.ams.usda.gov/AMSv1.0/FARMERSMARKETS**), which offers an annual count of operational farmers markets across the country. There are about 5,000 farmers markets in the United States. If you are accepted, you will probably have to pay membership dues. These can be between $75 and $150 up front, plus fees of $5 or so per day you set up.

A small-scale beginner who does not have enough products to justify the time spent at a farmers market could find a vendor to sell your beef for you, said A. Lee Meyer, an extension professor for sustainable agriculture of the University of Kentucky Department of Agricultural Economics. This partnership could work because the vendor will have another product to offer customers and you will benefit from an experienced seller's skills.

Community Supported Agriculture (CSAs)

Community Supported Agriculture is a system in which community members pay for "shares" of the food a farmer is expected to produce in the upcoming year. The money paid for these shares covers the cost of the farming expenses up front and allows farmers to do much of their marketing work during slow times of year. It ensures share holders get first choice of the types of food they want. For beef or dairy products, these share holders may be buying a stake in a particular animal. The downside to this for customers is they also share in the risks associated with farming, so if something goes wrong, such as the death of their animal, they may feel like they did not get their money's worth.

The concept of beef pooling, a variation on CSAs, has also been catching on. **Beef pooling**, also called "cow pooling," is another shareholder system in which one customer is in charge of coordinating orders for a group of people who buy in bulk. This person is the go-between for customers and producers, and it is his or her job to make sure all the customers in the pool get what they want. The beef pool coordinator can either ask everyone what cuts they want or simply take the initiative to divvy everything up so everyone gets a relatively equal share. These pools also ensure producers have an outlet for selling their meat.

Milk shares

Milk shares, also called herd shares, are arrangements that allow people to buy raw milk in some states (these shares are explicitly illegal in other states; check with state agriculture officials). The terms of these shares can vary by farm and could depend on the laws in your state. Often, customers pay a one-time fee for a share and also pay farmers a monthly boarding fee for the animal they own a share in. Customers often pay a deposit on the bottles that hold their milk. Herd shares entitle the owners to a set amount of milk, such as 1 gallon per week for the months cows are milked. Milk usually has to be picked up at the farm. Sometimes it can be picked up at a pre-arranged off-farm site. Sometimes, one person can pick up milk for a group of shareholders. Some farmers will buy back shares from customers who no longer want to participate; others allow shareholders to sell their share to another family. The terms of these shares vary by farm. There may be times of year when your animals are not producing milk. Make this clear to the customers.

CASE STUDY: ONLINE MARKETING

Ulla Kjarval
Co-founder: Sheepdog Print and Design
http://sheepdogpd.com

Many farmers have discovered the Internet is a powerful marketing tool that allows them to reach a broad number of people while spending little money. These farmers are using social media sites, websites that depend on user-generated content, as a way to connect with customers and to spread the word about their products. These sites are free and usually only require an e-mail address to join. Ulla Kjarval is a social media consultant who grew up on a grass-fed cattle farm in Meredith, New York, at the foothills of the Catskill Mountains. Her passion is not just the production of healthy food, but the preparation. She writes a food blog, **http://goldilocksfindsmanhattan.blogspot.com**, and with her sister Melkorka co-founded Sheepdog Print and Design (**http://sheepdogpd.com**), which offers Web design and photography and often works with farmers. Kjarval said these online marketing tools are cheap in that they do not cost money, but they do require a considerable time commitment. The benefit is each farmer can connect to potential customers and tell them about the parts of farming important to each customer.

"Everyone got into farming for different reasons, and everyone farms slightly differently," Kjarval said. "Talking about how you raise the animals, what kind of breed you have, all these things that seem really inane to a farmer are fascinating to someone who is not a farmer but is a potential customer."

Kjarval provided the following ways you can use social media to promote your farm's products:

BLOGS

A blog is a kind of online journal (the term blog is a shortened form of the words "Web log"). Writers of blogs write short entries as frequently as several times a day or once every few months. A blog would feature a series of articles displayed in order from newest to oldest. Two popular blog sites you can choose to host your blog are Blogger (**blogger.com**) and Blog.com (**blog.com**). Each farmer could tailor a blog to talk about his or her own interests. For example, if your passion is treating animals humanely, you can talk about that; or if you are more interested in the environmental benefits, that could be your focus. Some blogs are dedicated to cooking grass-fed products, rather than the growing. Or, you could even simply provide updates on your day-to-day activities. For example, one day you could let people know you are calving, and another day you could let people know you have new products available for sale. You could also use a blog as a feature on your own website.

"I think people should start out blogging," Kjarval said. "It is a great way to build confidence and understand what your narrative is. I think it is important to tell a story."

TWITTER

Twitter.com is a free site that lets users broadcast text messages called "tweets" that are 140 characters or less. Twitter might be a good choice for farmers who are not comfortable with their writing skills or who do not have the time to write whole blogs. You can set up your Twitter account so you can broadcast tweets from your phone; this would enable you to let people know when, for example, you were out in the field and about to move your cows to a fresh pasture. Users can decide if they want everyone to be able to see their tweets or if they only want certain users to view their tweets. Users who sign up for your tweets are called followers. Twitter also allows you to share links to websites, news stories, photos, and videos. The addresses to these sites (called URLs), are often longer than 140 characters, but you can convert them to a smaller number of characters at the site **http://tiny.cc**.

FACEBOOK

Users of this free site (**www.facebook.com**) create their own profile pages that highlight their personal information. Once you have a profile, you can become "friends" with other users and share information with each other by broadcasting updates to a news feed. This news feed is simply a list of all your friends' updates, usually shown with the newest updates first. In their news feed, users only see information posted by their friends. Farmers can use Facebook by starting pages devoted to their farms and potential customers can link up to them by either becoming a friend or becoming a fan. One way you could use Facebook is to update your status to let people know when you have fresh beef for sale.

"It's like subliminal messaging in a way," Kjarval said. "It is kind of like having an advertisement in people's lives."

YOUTUBE

YouTube (**www.youtube.com**) is a site that lets you upload and view videos. It is now inexpensive to film videos. Handheld Flip Video cameras cost less than $150. And the audience is huge; millions of people watch videos on this site. To upload content, register with the site. YouTube videos must be less than ten minutes and the file size must be no more than 2 gigabytes.

FLICKR

Flickr (**www.flickr.com**) is a site that lets people upload images and videos so they can be shared with other people. These images can be viewed on albums on the Flickr site or posted on other Web pages. You can upload images to Flickr from your computer or from your cell phone. Once they are uploaded, post them on other sites or e-mail them. Flickr asks you to categorize your photos so they can be found if users are searching for a particular topic, such as "cows" or "grass-fed cows." Flickr is owned by Yahoo!, so to create an account for the site you also need a Yahoo! account. Farmers could use it to post pictures from their farm or even provide slideshows that serve as a tour of the farm.

E-MAIL

Another simple way to increase sales is to collect e-mail addresses from everyone who buys from you. By doing this, you can reach out to potential customers the next time you have goods for sale. You might even let customers reserve what they want before it is ready.

New farmers could benefit from maintaining an online presence in ways other than marketing, Kjarval said. You will be able to stay in touch with farmers all over the country, and many of these farmers will be willing to share advice and support. YouTube has videos demonstrating many farming tasks you may not be familiar with. As an example, Kjarval said she knows a new goat farmer who watched video of a goat birth before his own animals delivered. There is even a nonprofit group, the AgChat Foundation (**www.agchat.org**), that helps farmers use social media to reach customers and other farmers. Use Twitter to participate by searching for #AgChat.

What Your Customers Need to Know

Even if you do everything right — turning the best animals or their milk into the best-tasting, healthiest grass-fed food and selling them for a profit to interested people — it will not matter if the customer ultimately does not enjoy the eating experience. Arm your customers with knowledge to ensure they store and cook your products the right ways. This section highlights tips your customers may need by incorporating information from the USDA Food Safety Inspection Service's "Beef...from Farm to Table" fact sheet.

Storing

Your customers may not know how long they can store meat in the refrigerator or how long it will keep in the freezer. Some tips, adapted from the USDA "Beef...from Farm to Table" fact sheet:

- You can store most beef cuts in the refrigerator at 40 degrees or colder for three to five days. Freeze it at 0 degrees for up to a year.

- Beef that has been cooked can be kept in a refrigerator for three or four days and frozen for three months or so.
- You can freeze beef in its original packaging or re-wrap it in freezer paper or aluminum foil. If you want to freeze it longer than that, keep it in the package and wrap the package with aluminum foil or freezer paper or place the package inside a plastic bag.

Thawing

This is a safety issue as much as anything. If you thaw out meat incorrectly, you can ruin it and make yourself sick. Meats should not be thawed at room temperature. You can cook frozen meat in an oven, or on a stove or grill, according to the USDA, but it will take longer. Frozen meat will not cook properly in a slow cooker.

The safe ways to thaw meat are in a refrigerator, in the microwave, or in a pan of cold water in a sink.

Refrigerator: This is a good option if you know in advance what you will be making. It takes several hours to thaw ground beef and steaks and can take a couple of days to thaw out a whole roast. After it thaws, keep it in the fridge for three to five days. If you change your mind in this time and do not cook it, you can re-freeze it.

Cold water: Leave it in the package and submerge it in cold water. Change the water every 30 minutes or so. It will take about an hour to defrost smaller cuts. Larger cuts could take two hours or more.

Microwave: Defrost in the microwave if you plan to cook right away. Parts of the meat could cook in the microwave as you thaw, allowing bacteria to survive unless you fully cook it right away.

Cooking

Many customers will not know the proper way to cook meat. It takes even more skill to cook many grass-fed cuts, which are leaner and often need to be cooked at lower temperatures and shorter cooking times. People often overcook meat, which makes them think it is tough. Marinating can help lean cuts.

The American Grassfed Association has recipes on its website, as do many grass-fed farmers. If you become good at cooking grass-fed products, you could also hold classes or cookouts on your farm to show your customers how it is done (this is also a nice way to bring in new customers).

The USDA "Beef… From Farm to Table" fact sheet also has a chart about cooking times and temperatures. If you are unsure if a dish is done, check the temperature with a meat thermometer.

APPROXIMATE BEEF COOKING TIMES °F

Type of Beef	Size	Cooking Method	Cooking Time	Internal Temperature
Rib Roast, bone in	4 to 6 lbs.	Roast 325°	23-25 min./lb.	Medium rare 145°
Rib Roast, boneless rolled	4 to 6 lbs.	Roast 325°	Add 5-8 min./lb. to times above	Same as above
Chuck Roast, Brisket	3 to 4 lbs.	Braise 325°	Braise 325°	Medium 160°
Round or Rump Roast	2 1/2 to 4 lbs.	Roast 325°	30-35 min./lb.	Medium rare 145°
Tenderloin, whole	4 to 6 lbs.	Roast 425°	45-60 min. total	Medium rare 145°
Steaks	3/4" thick	Broil/Grill	4-5 min. per side	Medium rare 145°
Stew or Shank Cross Cuts	1 to 1 1/2" thick	Cover with liquid; simmer	2 to 3 hours	Medium 160°
Short Ribs	4" long and 2" thick	Braise 325°	1 1/2 to 2 1/2 hours	Medium 160°

CASE STUDY: BARR FARMS

Adam Barr
Barr Farms
Rhodelia, Kentucky
adam.barr@gmail.com

Adam Barr is trying to do something that has not been done in generations: make a living completely off his family's farm in Kentucky. He believes product diversity will be the key to success. He offers grass-fed beef and chicken, plus mushrooms and vegetables, and he plans to offer nuts and berries. He sets up at farmers markets near downtown Louisville where there are few places to buy any vegetables, let alone those that are fresh and locally grown. His products are a novelty to many of his customers, and much of his sales pitch involves offering nutritional facts and cooking tips.

"A lot of it is education — customers do not know what to do with a rump roast," he said.

Many of his sales come through Community Supported Agriculture (CSA) arrangements, where customers pay a subscription fee before each season so they can receive food during the season. He has a CSA for vegetables, a CSA for eggs, and a meat CSA for cattle and chickens. His beef CSA runs in three-month increments. He collects an e-mail address from every customer so later he can send out a quarterly order form. The order form is a spread sheet Barr and his wife created that shows the price per pound of each cut. Customers have to fill in which products they want, and the spread sheet calculates the price total for them. Barr said he sells more individual cuts because customers do not have a place to store bulk orders. He usually gets about 15 to 20 people to order his beef each quarter. The orders are for as little as $5 worth of ground beef to as much as $150 worth of meat. He measures his success against the price per pound he would get at auction. He tries to sell his grass-finished beef for at least double what he would get per pound there. His father and uncle, who own the herd, earn $1 per pound of each animal's live weight before slaughter. For his efforts, Barr gets about 33 cents per pound.

At the farmers markets, he hands out a fact sheet detailing the nutritional content of his products. He does not seek certification under the National Organic Program because he cannot afford organic chicken feed. If he could obtain organic certification, the increased cost of the final product would put it out of reach for most of his customers. But he is listed under Certified Naturally Grown (**www.naturallygrown.org**), which adheres to similar principles as the National Organic Program but seeks to reduce much of the government paperwork and production costs that make the label prohibitive to many small farmers.

Barr's father and uncle, who both have full-time, off-farm jobs, each own about half of the 260 acres of farmland. Barr bought a couple of acres of the family farm in 2006. He helps manage the cattle grazing system and raises vegetables and chickens. The garden is a small-scale intensive production that uses double-dug raised beds, garden beds dug extra deep to stimulate root growth. His goal is to improve the pasture, little by little, using the animals as the tools and the end result being healthier soil. He moves the pastured egg-laying hens about 12 feet every day, letting them fertilize a new section of the pasture. You can see the difference in the parts where the chickens were — the grass is thick and bright and mixed with clover. He considers this high-quality area to be his finishing pasture. He is constantly experimenting with ways to improve his farm. A recent project involved hollowing out a bus he wants to use as a nesting area for chickens. They can lay their eggs in there and have shelter, and he can drag the bus around the pasture. A freshly grazed spot of pasture from this spring's first cow-calf grazing paddock was slated to become a bed for blueberries and raspberries.

The Barrs first harvested grass-finished beef in August 2008. Their herd includes red and black Angus and polled Hereford mixes. They sell about two-thirds of their calf crop at weaning. These calves are given some grain before they are sold. The other third are separated into a herd to be grass-finished. They target a finishing weight of at least 1,000 pounds, but they believe 1,200 pounds is better and gives more carcass yield. The cattle are usually 22 to 24 months old when they are finished. Barr and his partners use scales to monitor weight gains. They aim for at least 1 pound of gain per day for the last couple of months, and it usually takes 22 to 24 months to finish their cattle. It takes at least an extra year before the grass-finished animals are ready to sell, but the price can be up to three times as much as what they get for the weanlings. And the quality of the grass-finished beef gets better every year.

Conclusion

As more people become concerned about the way their food is produced, the demand for grass-fed cattle will continue to grow. Grass-fed meat and dairy foods are a niche product and will probably stay that way for the foreseeable future, but there will always be customers willing to pay more for food grown this way. It is healthier for people, it is stimulating for local economies, it is beneficial for farmland, and it is a better life for the animals that give us so much.

Grass-fed cattle farming is not easy. It requires a big investment in time: time to care for your animals, time to move them around your fields to sustain them and your pastures, and time to market them to paying customers. But many people enjoy this type of work. It is rewarding to be a part of something environmentally sustainable. Many farmers form bonds with their animals and their land as they become part of a beautiful natural cycle. Among a growing number of health-conscious and environmentally aware customers, small-scale farmers are heroes.

You no doubt have more questions, and will find yourself in many other farming situations when you are not sure what to do. The good news is many farmers are glad to pass on the wisdom they have collected, and there are many resources available for farmers with questions. *Many of these resources are listed in the Appendix.* The farmers and experts you come to trust will no doubt know of other sources that can tell you what you need to know.

If you want to be a grass-fed cattle farmer, you can do it. And you can have a lot of fun at it. You can even make money doing it. Take care of your animals and take care of your land, and your farm will take care of you.

• Appendix A •
Resources

Here are some resources you may find helpful in your farming endeavors.

General Information:

Cooperative extension offices

Extension offices are nationwide educational services staffed by experts called extension agents who provide information to farmers, children, small business owners, and others in rural and urban communities. Search a map of the United States for offices in your area at **www.csrees.usda.gov/Extension/index.html**.

United States Department of Agriculture

This federal agency oversees agriculture production and trade, offers information and assistance, and makes and enforces laws. Visit its website at **www.usda.gov**.

USDA agencies include:

- Farm Service Agency (FSA) (**www.fsa.usda.gov**)
- USDA Service Center. You can search by state at its website, **http://offices.sc.egov.usda.gov/locator/app**.
- To find Natural Resource Conservation Service offices, search at **http://offices.sc.egov.usda.gov/locator/app?agency=nrcs**.

American Grassfed Association

This organization has the most comprehensive standards about grass-fed cattle. The AGA website is **www.americangrassfed.org**.

National Sustainable Agriculture Information Service

ATTRA, the common name for the National Sustainable Agriculture Information Service, provides information about sustainable agriculture at **http://attra.ncat.org**.

The Stockman Grass Farmer

This is a respected trade publication with tips for success for farmers who want to make money using management-intensive grazine.

Eatwild

Eatwild (**http://eatwild.com**) offers a comprehensive site about the benefits of grass-fed farming. The website features research, advice, and lists of grass-fed farmers by state.

LocalHarvest

LocalHarvest (**www.localharvest.org**) offers a directory of small and organic farms and farmers markets.

Beginning Farmers

The Beginning Farmers website (**http://beginningfarmers.org**) has many resources for new farmers, including a list of land-link programs.

Legal Info

The website for the U.S. House of Representatives (**www.house.gov**) features a link to U.S. laws. The "government resources" section also links to state government sites where you can find more information about the state laws you must follow.

Business Planning

U.S. Small Business Administration (**www.sba.gov**) and county extension offices have checklists and guidelines for creating business plans.

The University of Kentucky College of Agriculture has budgeting examples and decision aid tools at **www.ca.uky.edu/agecon/index.php?p=565**.

The Oklahoma State University Department of Agricultural Economics (**http://agecon.okstate.edu/Quicken**) sells add-on software to use with the budgeting software Quicken to keep track of farm expenses and income.

Land-link programs

There are various online resources for land-link programs, which help farmers find land. One of the organizations you can reference is the International Farm Transition Network (**www.farmtransition.org**). The Farm Service Agency advertises properties for sale and gives beginning farmers first priority at **www.resales.usda.gov**.

Weather Planning

The National Climactic Data Center website (**http://cdo.ncdc.noaa.gov/cgi-bin/climatenormals/climatenormals.pl**) has a searchable database of temperature and precipitation data, including charts that break down the probable dates of first and last freezes in different areas of each state.

Marketing Help

California State University Chico has a website for grass-fed-cattle farmers that is a good resource for labeling info, including example labels, a guide to creating your own label, and links to marketing research. Visit its website at **www.csuchico.edu/agr/grassfedbeef**.

• Appendix B •

AGA Standards

The American Grassfed Association has the most comprehensive standards for raising grass-fed cattle. Section 3 of the AGA's standards includes rules for feeding and caring for animals. Printed here are excerpts of the standards as they appeared on the AGA website on Sept. 7, 2010. These are also subject to change. For the most up-to-date standards, visit **www.americangrassfed.org**.

3 GRASS-FED AND PASTURE-FINISHED STANDARDS

3.1 Forage Protocol

3.1.1: All livestock production must be pasture/grass/forage based.

3.1.2: Grass and forage shall be the feed source consumed for the lifetime of the ruminant animal, with the exception of milk consumed prior to weaning. The diet shall be derived solely from forage consisting of grass (annual and perennial), forbs (e.g.

Legumes, Brassica), browse, or cereal grain crops in the vegetative (pre-grain) state. Animals cannot be fed grain and must have continuous access to pasture.

3.1.3: Forage is defined as any herbaceous plant material that can be grazed or harvested for feeding, with the exception of grain.

3.2 Grazing, Confinement, and Stock-Piled Forages

3.2.1: AGA grass-fed and pasture-finished animals must be maintained at all times on range, pasture, or in paddocks with at least 75% forage cover or unbroken ground for their entire lives.

Note: The provisions of this standard do not preclude commonly used grazing practices such as high-density/low-frequency or strip grazing, when large numbers of animals may graze growing forages in small paddocks for short periods of time.

3.2.2: AGA grass-fed and pasture-finished animals must not be confined to a pen, feedlot, or other area where forages or crops are not grown during the growing season.

3.2.3: Exceptions to sections 3.2.1 and 3.2.2 are limited to emergencies that may threaten the safety and well being of the animals.

3.2.4: AGA grass-fed and pasture-finished animals may be fed hay, haylage, baleage, silage, crop residue without grain, and other roughage sources while on pasture.

3.2.5: AGA grass-fed and pasture-finished animals cannot be fed stock-piled forages in confinement for more than 30 days per calendar year.

3.2.6: Approved mineral and vitamin supplements may be provided free choice to adjust the animals' nutrient intakes and to correct deficiencies in their total diet.

3.2.7: Feeding of non-forage supplements is prohibited. If inadvertent exposure to non-forage feedstuffs occurs, the incident must be recorded as per 3.3.6.

3.2.8: Consumption of seeds naturally attached to herbage, forage, and browse is considered incidental and acceptable. Note: Oat hay and other cereal crops harvested after seed production are not permitted.

3.2.9: Deliberately waiting until a cereal grain crop has set seed before grazing or harvesting for stored forage is prohibited.

3.5 Animal Health and Welfare

3.5.1: All livestock-production methods used with AGA grass-fed ruminant animals must support humane animal welfare, handling, transport, and slaughter.

3.5.2: The producer must develop and maintain a written record of all vaccines, medications, or other substances used in their animal health-care program

3.5.3: AGA grass-fed ruminant animals must not be fed or injected with antibiotics.

3.5.4: Sick or injured animals must be treated to relieve their symptoms.

3.5.5: If prohibited medications are required for treatment, the animal must be identified, tracked, and recorded to demonstrate that it does not enter the AGA Grassfed Ruminant system.

Note: Provided the identification and tracking are adequate, the animal may still be kept with other animals that do qualify for the AGA Grassfed Ruminant system.

3.5.6: The producer must keep records of the purchase of any antibiotics. Antibiotic receipts and injection records must be available on demand to the certifying agency.

3.5.7: No hormones of any type may be administered to AGA grass-fed ruminant animals.

3.5.8: Livestock produced under these standards must not be fed any animal by-products at any time.

3.6 Animal Identification and Trace-Back

3.6.1: AGA grass-fed ruminant animals must be traceable by written record throughout their entire lives from birth to harvest to the farm or ranch from which they originated.

3.6.2: Each producer must develop and maintain an Animal Identification system to uniquely identify each animal and allow 48-hour trace-back.

3.6.3: Complete and up-to-date records must be maintained and specifically identify all animals raised and purchased that are sold or harvested as part of the AGA Grassfed Ruminant Program.

3.6.4: Complete and up-to-date records must be maintained to show the source of all purchased market animals brought onto farm or ranch.

3.6.5: Records must document the supplier of the purchased market animals raised them in accordance to AGA Grassfed and Pasture Finished Ruminant Standards.

3.6.6: Only market animals one year of age or younger may be brought into the AGA Certified Grassfed program by affidavit. *Note: AGA Grassfed Supplier Affidavit, Appendix A, must be used for purchased animals after January 1, 2010. All market animals brought onto a farm or ranch for the AGA Certified Grassfed Program must come from audited suppliers beginning January 1, 2011.*

3.6.7: Complete and up-to-date records must be maintained for all AGA market animals delivered for harvest. AGA Transfer of Livestock for Harvest documents are provided as a template.

3.6.8: All records are to be maintained for a minimum of 24 months after the animal is sold or harvested.

3.6.9: All required records must be in sufficient detail to demonstrate compliance with AGA standards to the certifying agency.

AMERICAN GRASSFED ASSOCIATION SUPPLEMENTS LIST

When supplements are used, it seems logical that these supplements should be looked upon as substitutes or replacements for the pasture that is not available at the time. Thus, the supplements should be nutritionally comparable in the major nutrient content of the forage being replaced. The nutrients considered should be uniformly available (nutritionally speaking) and probably include energy, fiber, starch, and protein. Because most pasture grasses, legumes, and mixtures contain 20 to 30 percent starch, it seems axiomatic that supplements should contain levels of starch along with a high level of highly digestible fiber. This rational indicates starch–fiber content is a better criterion to judge supplemental feedstuffs. The goal of any supplementation would be to not change the nutrient profile of the product produced (e.g. beef). The approach taken for this standard is to allow feeding of supplements on the basis of their Nonfiberous Carbohydrate Content (NFC). Supplements with an NFC percent of 30 percent or less will be allowed under this standard.

The following list of approved supplements is not an exclusive list but lists supplements that have been approved by the AGA to date. The AGA Certification Committee may review and amend this list periodically. Supplements not listed below must be approved in advance by AGA's nutritional supplements committee. Supplements that have an adverse effect on the nutritional quality or have negative health benefits on the animals will not be allowed.

AGA APPROVED SUPPLEMENTS LIST — TIER 2

- Cottonseed Hulls or Cottonseed-Hull Pellets
- Cottonseed Meal mechanically or solvent extracted
- Peanut Hulls or Peanut-Hull Pellets
- Peanut meal
- Rice Hulls or Rice-Hull Pellets
- Alfalfa Cubes or Pellets (17% Protein)
- Corn Cobs
- Oat Hulls or Oat-Hull Pellets
- Oat Silage (dough stage)
- Corn silage (no grain)
- Soy Hulls or Soy-Hull Pellets
- Beet Pulp
- Flax Seed or Flax-Seed Meal
- Safflower Seed Meal
- Corn Gluten
- Wheat Bran
- Sunflower Meal, mechanically or solvent extracted

ADDITIONAL APPROVED SUPPLEMENTS FOR TIER 3

All the supplements listed above are approved for Tier 3 as well as the supplements listed below.

- Brewer's Grain
- Distiller's Grain

AMERICAN GRASSFED ASSOCIATION AGA BANNED FEEDSTUFFS LIST

The following list of banned feedstuffs is not an exclusive list. The AGA Certification Committee may review and amend this list periodically.

- Corn
- Cereal Grains
- Urea and other non-protein nitrogen sources
- Milk replacer containing antibiotics, growth promoters, and/or any animal by-products aside from milk protein
- Animal by-products
- Antibiotics
- Hormones

Glossary

Aging: letting beef hang in the freezer for several days so natural enzymes can break down connective tissues that make beef tough

Animal Unit Equivalent: a measurement defined as a 1,000-pound animal, the average weight of a mature beef cow

Annuals: plants that grow and die within a one-year time period

Artificial insemination: breeding cattle using harvested semen as opposed to mating a cow with a live bull

Augur: a drill-like tool attached to a tractor or a skid steer

Backgrounder: a type of beef cattle operation where farmers buy weaned calves and fatten them up cheaply before sending them to feedlot operations

Beef pooling: a shareholder system in which one customer is in charge of coordinating orders for a group of people who buy in bulk; these pools ensure producers have an outlet for selling their meat

Biosecurity: preventative steps farmers can take to keep new illnesses from infecting animals

Breeder auctions: auctions held for animals with certain genetics or for certain breeds

Broadcast seeder: a machine used to apply fertilizer or seeds by spraying or casting seeds outward in many directions

Bucket system: a system of milking where vacuum tubes pump milk into buckets and the milk is poured from the buckets to a holding container

Bulk tank: a stainless steel refrigeration unit that keeps milk cool until it is collected

Bull: an uncastrated male cow

Calf jack: a tool that has a looping chain placed around a calf's legs and is used to pull out a calf 1 inch at a time

Carrying capacity: the stocking rate your pasture can support

Community Supported Agriculture: a system in which community members pay for "shares" of the food a farmer is expected to produce in the upcoming year

Continuous grazing: a system where animals are kept in the same pasture area all the time

Cooperative extension: a nationwide educational service offered by experts called extension agents who provide information to farmers, children, small business owners, and others in communities

Cooperatives: groups of farmers who share resources and market their collective products under one brand name

Co-packer: a company that packages products to a farmer's specifications

Creep grazing: a grazing system where unweaned calves are allowed to graze pasture ahead of their mothers

Crossbreeding: mating cows with bulls of different breeds

Cud: pieces of chewed and regurgitated food

Culling: selling animals you do not want to raise any more; animals you do not want to use anymore for meat or milk

Custom livestock haulers: contractors who will transport your animals on or off your property for a fee

Custom-exempt plants: at these plants, only the facilities (not the animals) are inspected; the plants are only allowed to process animals for the use of the owner and non-paying guests

Desertification: thinning of plant cover due to weakened soil productivity

Destocking: culling cows

Direct marketing: selling meat or milk products directly to the consumers

Disks: devices pulled behind tractors that have adjustable blades to break up the top layer of soil

Drought: a prolonged period of dry weather

Dry matter: the amount of feed in plant material after all water is removed

Estrous cycle: the cycle from one period of heat to the next

Extension agent: a person who provides educational information to farmers and others in communities through a cooperative extension office

Feedlot: areas where cattle are gathered to be fattened on high-grain diets before slaughter

Fenceline weaning: calves are separated from their mothers into an adjoining pasture where they can still see, hear, and smell their mothers; this is considered a weaning strategy used to reduce stress

First-calf heifers: a cow that has produced just one calf

Flight zone: a radius around the cow that the animal wants people to steer clear of

Forage: food grazing animals eat

Frost seeding: a technique where farmers spread seeds onto pasture in late winter or early spring while the ground is still frozen

Grafting: the process of introducing calves to a nurse cow

Grass finish: using an all-pasture diet to fatten a beef animal so it can be used for meat

Graziers: farmers who manage grazing animals

Hanging weight: the number of pounds beef weighs while hanging on a rail in the cooler

Heifer: a cow older than 1 year that has not had a calf

Herd sire: the bull that contributes most of the genetics to the cows on your farm

Heterosis: bringing out the best in both breeds; also called hybrid vigor

Homeopathy: a system of treatment using natural substances instead of traditional treatment options

Homogenized: when a machine is used to break up fat globules into small pieces so the cream will not rise to the top

Humus: organic, decayed matter that houses microorganisms; it is useful to use in gardening and to improve soil

Irrigation: artificially applying water to pasture and fields

Leader-follower grazing: a grazing system where animals with higher nutritional needs (calves and lactation cows for example) are given first access to the best grasses

Live weight: the amount an animal weighs just before it is slaughtered

Milk letdown: the process of releasing milk into the udder

Milk shares: an arrangement that allows people to buy raw milk

Mob feeding: keeping groups of four to ten calves in a paddock separate from their mothers and using a large barrel with nipples on it to feed the calves

Nurse cow: a cow used to nurse her own calf and other dairy calves

On the hoof: the practice of selling the animal or shares of the animal when it is still alive

Open cows: non-pregnant cows

Open heifer: a heifer that fails to breed

Organic plan: a plan that describes how a farmer will run an organic farm, including what types of feed will be used and where the substances will come from

Overgrazing: allowing animals to graze plants faster than plants can regrow

Paddock: a subdivided section of pasture in a rotational grazing system

Pasteurization: the process of heating milk to kill bacteria

Perennials: plants that grow back from the same root system each year

Plate cooler: a device that pre-cools milk before it gets to the storage tank

Polled: having no horns

Post-hole digger: a clamshell-shaped tool used to dig holes for posts to be placed in

Progeny: future offspring

Prolapsed: when the uterus is on the outside of the birth canal

Raw milk: milk that has not been pasteurized

Rejuvenating: the process of adding new plants to an existing pasture

Relative feed value: this is calculated by the amount of energy and protein present in the hay to determine its nutritional content related to grass

Replacement heifers: female cows used for breeding and/or milking

Re-seeding: the process of tearing up a plot of land and starting over so farmers have control over the dominant species of plants

Riding: when a female mounts other females; this may be a sign of heat

Riparian area: the transitional area of land near water sources

Rotational grazing: dividing pasture into sections and rotating herds from one section to the next

Ruminants: even-toed, hooved animals that chew cut; includes cattle, bison, goats, and sheep

Sacrifice area: a place where farmers keep animals for an extended period of time when the ground is vulnerable, such as during dry weather or in winter

Salvage value: the price you receive for selling an older or unproductive cow

Seasonal calving: putting bulls and breeding-age females together for a predetermined period of time to ensure that all cows will get pregnant and deliver in the same time frame

Seasonal dairy: a dairy farm that delivers its calves all in one season

Seed drill: a planting drill that has blades that cut the soil and drop seeds into the cuts to ensure good contact with the soil

Set stocking: a strategy of turning your finishing animals onto a lush piece of pasture and let them eat until they are ready for slaughter

Silage: fermented, moist forage made from almost any green, growing plant

Skid steer: a useful piece of equipment that can do a lot of the heavy lifting required on the farm

Springing: when the vulva and tail head may become loose and jiggle on a pregnant cow; this usually occurs ten days before she gives birth

Stanchions: devices attached to a cow's head to keep it still during milking

Standing heat: a period of heat where cows allow other cattle, including females, to mount them as they stand

Steer: a castrated male calf unsuitable to use as a bull

Stocker cattle: cattle in between the weaning phase and the finishing stage

Stocking rate: the number of AUE per acre a pasture can support

Strip grazing: a method of grazing where farmers temporarily move an electric fence back a little each day to introduce new forage

Tie-stalls: tools used to hold cows during milking

Trim: meat leftover after the butcher removes the best cuts

Vegetative stage: the growth stage in a plant's life cycle

Weanling: an animal that has just been weaned

Yearling: an animal between 1 and 2 years old

Bibliography

"About the Breed." Dutch Belted Cattle Association of America. **www.dutchbelted.com/About%20the%20Breed.html**. Accessed on Sept. 6, 2010.

"About Us." International Veterinary Acupuncture Society. **www.ivas.org**. Accessed on Aug. 27, 2010.

"About Us." Real Raw Milk Facts. **www.realrawmilkfacts.com/about-us**. Accessed on Sept. 20, 2010.

"Access to Pasture (Livestock); Proposed Rule." National Organic Program. U.S. Department of Agriculture Agricultural Marketing Service, 7 CFR Part 205. October 24, 2008. **www.ams.usda.gov/AMSv1.0/getfile?dDocName=STELPRDC5073426&acct=noprulemaking**. Accessed on March 11, 2010.

Adam, Katherine L. "Directory of Organic Seed Suppliers." ATTRA – National Sustainable Agriculture Information Service. **http://attra.ncat.org/attra-pub/organic_seed**. Accessed on Sept. 6, 2010.

Agri-Dynamics. **www.agri-dynamics.com**. Accessed on Aug. 31, 2010.

American Chianina Association. **www.chicattle.org**. Accessed on Sept. 5, 2010.

American Criollo Beef Association. **www.leanandtenderbeef.com/about-criollo-cattle**. Accessed on Sept. 6, 2010.

American Criollo Beef Association. **www.leanandtenderbeef.com/About-Criollo-Cattle/Criollo-Cattle**. Accessed on Sept. 6, 2010.

American Galloway Breeders Association. **http://americangalloway.com**. Accessed on Aug. 30, 2010.

American Highland Cattle Association. **www.highlandcattleusa.org**. Accessed on Sept. 5, 2010.

American-International Charolais Association. **www.charolaisusa.com**. Accessed on Sept. 5, 2010.

American Livestock Breeds Conservancy. **www.albc-usa.org**. Accessed on Sept. 27, 2010.

American Milking Shorthorn Society. **www.milkingshorthorn.com.** Accessed on Sept. 5, 2010.

Anderson, Neil. "Dehorning Of Calves." The Cattle Site. January 2009. **www.thecattlesite.com/articles/2261/dehorning-of-calves**. Accessed on June 8, 2010.

Andrae, John and Hancock, Dennis. "Common Grazing Methods and Some Specific Farm Applications." July 16, 2010. The University of Georgia, College of Agricultural & Environmental Sciences. **www.caes.uga.edu/topics/sustainag/grazing/grazingsysdes/commongraz.html**. Accessed on Sept. 1, 2010.

Ayrshires Cattle Society. **www.ayrshirescs.org**. Accessed on Aug. 31, 2010.

Baker, Barton. "Creep Grazing." West Virginia University Extension Service. Feb. 1996. **www.caf.wvu.edu/~forage/creep_grazing.htm**. Accessed on Sept. 4, 2010.

Balkcom, Carrie. Executive director, American Grassfed Association. Personal interview on Feb. 26.

Ballard, Ed. "Extending the Grazing Season." University of Illinois Extension: Illini PastureNET. Oct. 31, 2005. **www.livestocktrail.uiuc.edu/pasturenet/paperDisplay.cfm?ContentID=8146**. Accessed on March 19, 2010.

Barnhart, Stephen K. "Stockpiled forages: A way to extend the grazing season." Iowa State University, University Extension. Nov. 1998. **www.extension.iastate.edu/Publications/PM1772.pdf**. Accessed on May 4, 2010.

Barr, Adam. Personal interview on April 16, 2010.

Bartholomew, Henry, M. "Getting Started Grazing." Ohio State University. **http://ohioline.osu.edu/gsg**. Accessed on March 19, 2010.

Beard, F. Richard et al. "Maintenance of Wheelmove Irrigation Systems." Utah State University Extension. August 2000. **www.msuextension.org/ruralliving/Dream/PDF/maintainwheelline.pdf**. Accessed on Sept. 4, 2010.

"Beef…From Farm to Table." Food Safety Inspection Service. U.S. Department of Agriculture. Oct. 19, 2009. **www.fsis.usda.gov/factsheets/beef_from_farm_to_table/index.asp**. Accessed on Sept. 22, 2010.

"Beef Sire Selection for Cattle Genetic Improvement Program." University of Kentucky, College of Agriculture. Updated February 9, 2007. **www.uky.edu/Ag/AnimalSciences/pubs/beefsireselectionguidelines.pdf**. Jan. 9, 2006.

Beegle, Douglas. "Soil fertility management for forage crops, Pre-establishment." Pennsylvania State University, College of Agricultural Sciences. **http://pubs.cas.psu.edu/FreePubs/pdfs/uc096.pdf**. Accessed on March 7.

Beetz, Alice and Rinehart, Lee. "Pastures: Sustainable Management." ATTRA – National Sustainable Agriculture Information Service. p.9, 12. 2006. **http://attra.ncat.org/attra-pub/PDF/sustpast.pdf**. Accessed on Aug. 31, 2010.

Bennett, Mike. Personal interviews Feb. 11, 2010, June 8, 2010, and Sept. 6, 2010.

Berry, Wendell. *Bringing It to the Table*. Counterpoint Press. Berkeley, Calif. 2009.

Bowman, Gary L. and Shulaw, William P. "On-Farm Biosecurity: Traffic Control and Sanitation." Ohio State University Extension Fact Sheet. June 2001. **http://ohioline.osu.edu/vme-fact/0006.html**. Accessed on Oct. 1, 2010.

"Breeds of Livestock." Oklahoma State University. **www.ansi.okstate.edu/breeds/cattle**. Accessed on Oct. 1, 2010.

"Brucellosis." Department of Health and Human Services: Centers for Disease Control and Prevention. National Center for Immunization and Respiratory Diseases: Division of Bacterial Diseases. Dec. 7, 2007. **www.cdc.gov/ncidod/dbmd/diseaseinfo/brucellosis_g.htm**. Accessed on Aug. 31, 2010.

"BSE (Bovine Spongiform Encephalopathy, or Mad Cow Disease)." Department of Health and Human Services: Centers for Disease Control and Prevention. Aug. 26, 2010. **www.cdc.gov/ncidod/dvrd/bse**. Accessed on Sept. 6, 2010.

"Budgets/Decision Aids." University of Kentucky, College of Agriculture. **www.ca.uky.edu/agecon/index.php?p=565**. Accessed on Sept. 6, 2010.

Bullock, Darrh. "Using EPDs, Expected Progeny Differences." Cooperative Extension Service. University of Kentucky, College of Agriculture. Sept. 1993. **http://www.ca.uky.edu/agc/pubs/asc/asc141/asc141.htm**.

"Bureau of Reclamation – About Us." U.S. Bureau of Reclamation. Nov. 16, 2009. **www.usbr.gov/main/about**. Accessed on March 6.

Cadwallader, Tom and Cosgrove, Dennis. "Fencing Systems for Rotational Grazing: Bracing." University of Wisconsin–Extension. March 27, 2008. **www.uwrf.edu/grazing/bracing.pdf**. Accessed on Oct. 1, 2010.

Calf Cozy. **www.calfcozy.com**. Accessed on Sept. 6, 2010.

Cassell, Bennet and McAllister, Jack. "Dairy Crossbreeding: Why and How." Cooperative Extension System. May 05, 2010. **www.extension.org/pages/Dairy_Crossbreeding:_Why_and_How**. Accessed on Sept. 6, 2010.

"Certification in 5 Steps." American Grassfed Association. **www.americangrassfed.org/wp-content/uploads/2009/02/certification_step_by_step_for_website[1].pdf**. Accessed on Aug. 30, 2010.

"Certified Laboratories." National Forage Testing Association. **www.foragetesting.org/index.php?page=certified_labs**. Accessed on Aug. 26, 2010.

"Characteristics of Criollo Cattle." American Devon Cattle Association. **www.americandevon.com**. Accessed on Sept. 5, 2010.

"Climate of Nevada." Western Regional Climate Center. National Climate Service. **www.wrcc.dri.edu/narratives/NEVADA.htm.**

"Cooperatives In the Dairy Industry." U.S. Department of Agriculture. Sept. 2005. **www.rurdev.usda.gov/RBS/pub/cir116.pdf**. Accessed on June 28.

Crary, Vince. "Irrigated Dairy Pasture Soil Water Profile Research Project." University of Minnesota Extension. Published in *Dairy Star*. June 20, 2008. **www1.extension.umn.edu/dairy/grazing-systems/irrigated-dairy-pasture-research**. Accessed on Sept. 4, 2010.

Crystal Creek Inc. **www.crystalcreeknatural.com**. Accessed on Aug. 31, 2010.

Dimitri, Carolyn, and Oberholtzer, Lydia. "Marketing U.S. Organic Foods: Recent Trends From Farms to Consumers." Economic Information Bulletin No. 58. U.S. Dept. of Agriculture, Economic Research Service. September 2009. **www.ers.usda.gov/publications/eib58.**

Duckett, Susan K. "Understanding Factors Affecting Meat Quality." Clemson University. **www.americangrassfed.org/wp-content/uploads/Meat%20 Quality-Susan%20Duckett.pdf**. Accessed on Oct. 1, 2010.

Dunn, J.L. et al. "Identification of optimal ranges in ribeye area for portion cutting of beef steaks." Journal of Animal Science. April 2000. p.966. Cited in "Packing Industry Innovations As We Move to the Future?" University of Nebraska – Lincoln. 2001. **http://digitalcommons.unl.edu/cgi/viewcontent.cgi?article=1102&context=rangebeefcowsymp**. Accessed on Oct. 1, 2010.

Drake, Daniel J. and Oltjen, James. "Intensively Managed Rotational Grazing Systems for Irrigated Pasture."California Cooperative Extension California Ranchers' Management Guide. 1994, **http://ag.arizona.edu/arec/pubs/rmg/1%20rangelandmanagement/6%20rotategrazesystems94.pdf**. Accessed on Oct. 1, 2010.

Dr. Temple Grandin's Web Page. **www.grandin.com**. Accessed on March 22.

Ekarius, Carol. *Small-Scale Livestock Farming*. Storey Publishing. North Adams, Mass. 1999.

"Electric Fencing for Serious Graziers." Missouri Natural Resources Conservation Service. 2005. **www.mo.nrcs.usda.gov/news/pubs_download/out/MO%20NRCS%20Electric%20Fencing_low.pdf**. Accessed on Aug. 31, 2010.

Estabrook, Barry. "The Need for Custom Slaughter." The Atlantic. Jan. 21, 2010. **www.theatlantic.com/food/archive/2010/01/the-need-for-custom-slaughter/33904**. Accessed on July 19, 2010.

Family Milk Cow Dairy Supply Store. **www.familymilkcow.com**. Accessed on Aug. 31, 2010.

Farm Service Agency. U.S. Department of Agriculture. **www.fsa.usda.gov**.

"Farmers Markets." Agricultural Marketing Service. U.S. Department of Agriculture. **www.ams.usda.gov/AMSv1.0/FARMERSMARKETS**. Accessed on Sept. 6, 2010.

Fiedler, Jim. Personal interview on July 13, 2010.

Flitner, Dan. Personal interview on March 29, 2010.

"Frequently Asked Questions." Certified Naturally Grown. **www.naturallygrown.org/about-cng/frequently-asked-questions**. Accessed on April 16, 2010.

Gay, Susan and Heidel, Richard D. "Constructing High-tensile Wire Fences." Virginia Tech, Virginia Cooperative Extension. May 1, 2009. **http://pubs.ext.vt.edu/442/442-132/442-132.html**. Accessed on March 28, 2010.

Gegner, Lance E. "Organic Alternatives to Treated Lumber." ATTRA – National Sustainable Agriculture Information Service. 2002. **http://attra.ncat.org/attra-pub/lumber.html#usda**. Accessed on March 15, 2010.

Glengarry Cheesemaking and Dairy Supply. **www.glengarrycheesemaking.on.ca**. Accessed on Sept. 6, 2010.

Go-To Tanks. **www.gototanks.com**. Accessed on Aug. 31, 2010.

"Grass Fed Marketing Claim Standards." Agricultural Marketing Service. U.S. Department of Agriculture. Oct. 16, 2007. **www.ams.usda.gov/AMSv1.0/ams.**

fetchTemplateData.do?template=TemplateN&navID=GrassFedMarketingClaimStandards&rightNav1=GrassFedMarketingClaimStandards&topNav=&leftNav=GradingCertificationandVerfication&page=GrassFedMarketingClaims&resultType=&acct=lss. Accessed on June 28, 2010.

"The Grass-Fed Revolution." *Time*. June 11, 2006. **www.time.com/time/magazine/article/0,9171,1200759-3,00.html**. Accessed on Feb. 8, 2010.

"Grassed & Pasture Finished Ruminant Standards." American Grassfed Association. July 2009. **www.americangrassfed.org/wp-content/uploads/AGA%20Grassfed%20Standards%207-17-09.pdf**. Accessed on Sept. 6, 2010.

"Graziers Glossary." The Stockman Grass Farmer. **www.stockmangrassfarmer.net/FAQ.html**. Accessed on Feb. 22, 2010.

Greiner, Scott P. "Understanding Expected Progeny Differences (EPDs)." Virginia Tech University. May 1, 2009. **http://pubs.ext.vt.edu/400/400-804/400-804.html**. Accessed on Sept. 6, 2010.

"A Guide to Federal Food Labeling Requirements for Meat and Poultry Products." Food Safety Inspection Service. U.S. Department of Agriculture. **www.fsis.usda.gov/PDF/Labeling_Requirements_Guide.pdf**. Accessed on Sept. 6, 2010.

Hancock, Dennis. "Living With Johnsongrass." University of Georgia. July 2008. **www.caes.uga.edu/commodities/fieldcrops/forages/Ga_Cat_Arc/2008/GC0807.pdf**. Accessed on Sept. 4, 2010.

Halich, Greg. "Profitability Evaluation for Pasture-Based Finishing Production Scenarios." University of Kentucky, College of Agriculture. Pasture Based Beef for Local Markets Workshops. Feb. 18, 2010. **www.uky.edu/Ag/pasturebeef/pubs/halich-session2.pdf**.

Halich, Greg. Assistant extension professor, University of Kentucky, Department of Agricultural Economics. Personal interview on March 15, 2010.

Harris, Will. Personal interview on March 9, 2010.

Hasheider, Philip. *How to Raise Cattle*. Voyageur Press. Minneapolis, Minn. 2007.

The Hay Barn. **www.haybarn.com**. Accessed on Aug. 27, 2010.

Henning, Jimmy et al. "Rotational Grazing." University of Kentucky. **www.ca.uky.edu/agc/pubs/id/id143/id143.htm**. Accessed on Sept. 6, 2010.

"History of American Milking Devon Cattle." American Milking Devon Cattle Association. **www.milkingdevons.org/hist.html**. Accessed on Sept. 5, 2010.

"History of the Montbéliarde breed." Organisme de Selecion de la Race Montbeliarde. **www.montbeliarde.org/historique-en.php**. Accessed on Sept. 6, 2010.

"History of Shorthorns." American Shorthorn Association. **www.shorthorn.org/breedinfo/about/history_about_breedinfo.html**. Accessed on Sept. 5, 2010.

Horner, Joe et al. "Missouri 150-Cow Grazing Dairy Model." University of Missouri Extension. Sept. 2007. **http://agebb.missouri.edu/dairy/dairylinks/150-CowGrazing.pdf**. Accessed on March 15, 2010.

Horner, Joe. Extension associate of agricultural economics, University of Missouri, College of Agriculture, Food, and Natural Resources Personal interview on March 17, 2010.

"How a Corn Plant Develops." Iowa State University, University Extension. June 1993. **www.extension.iastate.edu/hancock/info/corn.htm**. Accessed on Sept. 21, 2010.

Huffstutter, P.J. "Raw Food Raid Highlights a Hunger." *Los Angeles Times*. July 25, 2010. Based on National Conference of State Legislatures research. **www.latimes.com/business/la-fi-raw-food-raid-20100725,0,4951907.story**. Accessed on Sept. 6, 2010.

Hyde, Lauren. "Limousin Breeders Tackle Temperament – Genetic trend shows power of selection." North American Limousin Foundation. **www.nalf.org/pdf/2010/aug19/tackletemperament.pdf**. Accessed on Aug. 30, 2010.

"Irrigation and Water Use: Glossary." U.S. Department of Agriculture. Economic Research Service. October 26, 2004. **www.ers.usda.gov/briefing/wateruse/glossary.htm**. Accessed on March 2, 2010.

Jarvis, Michael and Cox, Billy. Agricultural Marketing Service. "USDA Issues Final Rule on Organic Access to Pasture." U.S. Department of Agriculture. Feb. 12, 2010. **www.ams.usda.gov/AMSv1.0/ams.fetchTemplateData.do?template=TemplateU&navID=&page=Newsroom&resultType=Details&dDocName=STELPRDC5082658&dID=126904&wf=false&description=USDA+Issues+Final+Rule+on+Organic+Access+to+Pasture+&topNav=Newsroom&leftNav**. Accessed on June 13, 2010.

Kellogg, Wayne. "Body Condition Scoring With Dairy Cattle." University of Arkansas, Division of Agriculture. p.3–4. **www.uaex.edu/Other_Areas/publications/PDF/FSA-4008.pdf**. Accessed on April 7, 2010.

"Kentucky Hay For Sale Listings." Internet Hay Exchange. **www.hayexchange.com/ky.php**. Accessed on Aug. 27, 2010.

Kilde, Rebecca S. and Kleinschmit, Martin. "Summer Calving: A Practice to Improve Profits." North Central Initiative for Small Farm Profitability Center for Rural Affairs. **http://agmarketing.extension.psu.edu/begfrmrs/OptStratSmlFrms/AltProdPractices/SumrCalving.pdf**. Accessed on Aug. 31, 2010.

Kirkpatrick, F. David. "Management Of The Beef Bull." University of Tennessee, Animal Science Department. 2004. **http://animalscience.ag.utk.edu/beef/pdf/ManagementoftheBeefBull-FDK-2004.pdf**. Accessed on April 11, 2010.

Kjarval, Ulla. Personal interview on June 1, 2010.

Kopcha, Michelle. "Rabies: A Rare But Important Disease." Michigan Dairy Review. Michigan State University. Jan. 2010. **www.msu.edu/~mdr/vol15no1/vol15no1.pdf**. Accessed on Aug. 27, 2010.

Kramer, Fred D. and Johnson, Keith D. "Producing Emergency or Supplemental Forage for Livestock." Purdue University, Agronomy Department. 2007. **www.agry.purdue.edu/ext/forages/publications/ay263.htm**. Accessed on Sept. 5, 2010.

Kriegl, Tom. "Summary of Economic Studies of Organic Dairy Farming in Wisconsin, New England, and Quebec." University of Wisconsin-Madison. March 20, 2006. **http://cdp.wisc.edu/pdf/Organic%20Econ%20in%20N%20E,%20Q,%20&%20W1.pdf**. Accessed on April 10, 2010.

Lacefield, Garry et al. "Establishing Forage Crops." University of Kentucky, College of Agriculture. July 2003. **www.ca.uky.edu/agc/pubs/agr/agr64/agr64.pdf**. Accessed on March 18, 2010.

Lardy, Greg et al. "Spring versus Summer Calving for the Nebraska Sandhills: Production Characteristics" University of Nebraska–Lincoln, Animal Science Department. 1988. **http://digitalcommons.unl.edu/cgi/viewcontent.cgi?article=1348&context=animalscinbcr**. Accessed on Aug. 31, 2010.

Launchbaugh, Karen. "Forage Production and Carrying Capacity: Guidelines for Setting a Proper Stocking Rate." University of Idaho, College of Natural Resources. **www.cnr.uidaho.edu/what-is-range/Curriculum/MOD3/Stocking-rate-guidelines.pdf**. Accessed on Aug. 19, 2010.

LeClair, W.R. Personal interview on April 3, 2010.

Lee, C.D et al. "Grain and Forage Crop Guide for Kentucky." University of Kentucky, College of Agriculture. Cooperative Extension Service. May 2007. **www.ca.uky.edu/agc/pubs/agr/agr18/agr18.pdf**.

"Leptospirosis and Your Pet." Department of Health and Human Services: Centers for Disease Control and Prevention. Oct. 12, 2005. **www.cdc.gov/ncidod/dbmd/diseaseinfo/leptospirosis_g_pet.htm**. Accessed on Aug. 30, 2010.

Little, Carl. "Don't Let Your Herd Share Agreement Land You In Court." Real Milk Articles. **www.realmilk.com/herd-share-legalities.html**. Accessed on June 28.

"Livestock and Seed Auditing Services." Agriculture and Marketing Service. U.S. Department of Agriculture. **www.ams.usda.gov/ARCaudits**. Accessed on Aug. 31, 2010.

"Livestock Handling Facilities Designs." Illinois Grazing Manual. U.S. Department of Agriculture. Dec. 2006. **ftp://ftp-fc.sc.egov.usda.gov/IL/grazing/LiveHandFacilDesigns.pdf**. Accessed on Sept. 6, 2010.

Lozier, John et al. "Growing and Selling Pasture-Finished Beef: Results of a Nationwide Survey." Journal of Sustainable Agriculture. Vol. 25, Issue 2, 2004. p.93–112. **www.caf.wvu.edu/~forage/PFBSurvey.pdf**. Accessed on Sept. 6, 2010.

Mangione, David A. "Scoring Cows Can Improve Profits." Ohio State University, Department of Animal Sciences. **http://ohioline.osu.edu/l292/index.html**. Accessed on Sept. 6, 2010.

McBride, William D. and Greene, Catherine. "Characteristics, Costs, and Issues for Organic Dairy Farming." U.S. Department of Agriculture, Economic Research Service. 2009. **www.ers.usda.gov/Publications/ERR82/ERR82.pdf**. Accessed on Feb. 21, 2010.

"Member State Farmers Market Organizations." Farmers Market Coalition. **http://farmersmarketcoalition.org/state-associations**. Accessed on Sept. 6, 2010.

Meyer, A. Lee. Extension professor for sustainable agriculture, University of Kentucky, Department of Agricultural Economics. Personal interview on March 19, 2006.

"Minnesota — Climate." City-Data.com. **www.city-data.com/states/Minnesota-Climate.html**.

Morrison, E. M. "Two Minnesota co-ops are betting that dairy products from pasture-fed cows will be the next niche." Agricultural Utilization Research Institute. April 2001. **www.auri.org/agnews-section.php?sid=422&agnid=79**. Accessed on June 10, 2010.

Morrison, Rod. Personal interview on Feb. 16, 2010.

Morrow, Ron et al. "Paddock Design, Fencing, and Water Systems for Controlled Grazing." ATTRA – National Sustainable Agriculture Information Service. **http://attra.ncat.org/attra-pub/paddock.html**. 2005, 2009. Accessed on March 11, 2010.

Morrow, Ron et al. "Practical Use Leader-Follower Grazing Systems." University of Missouri-Columbia, College of Agriculture, Food and Natural Resources. Forage Systems Research Center. **http://aes.missouri.edu/fsrc/news/archives/nl94v3n1.stm**. Accessed on Sept. 6, 2010.

"Multistate Outbreak of E. coli O157:H7 Infections Associated with Beef from Fairbank Farms." Department of Health and Human Services, Centers for Disease Control and Prevention. Nov. 24, 2009. **www.cdc.gov/ecoli/2009**. Accessed on Sept. 6, 2010.

Nation, Allan. "Summer annuals can help keep animal performance high and produce tender meat." The Stockman Grass Farmer. **www.stockmangrassfarmer.net/cgi-bin/page.cgi?id=647**. Accessed on Sept. 4, 2010.

Nation, Allan. "Tips on Irrigated Pasture Management." The Stockman Grass Farmer. **www.stockmangrassfarmer.net/cgi-bin/page.cgi?id=589**. Accessed on Sept. 4, 2010.

National Organic Program. **www.ams.usda.gov/AMSv1.0/NOP**. Accessed on Sept. 6, 2010.

"Glossary of Terms." USDA Natural Resources Conservation Service. **http://soils.usda.gov/sqi/concepts/glossary.html**. Accessed on Aug. 31, 2010.

"NCAT's Organic Livestock Workbook." National Center for Appropriate Technology. ATTRA. Feb. 2004. **www.attra.org/attra-pub/PDF/livestockworkbook.pdf**. Accessed on Feb. 17, 2010.

Neel, J.P.S. et al. "Effects of winter stocker growth rate and finishing system on: I. Animal performance and carcass characteristics." Journal of Animal Science. 85:2007. p.2012–2018. **http://jas.fass.org/cgi/content/short/85/8/2012.**

Nelson, Melissa G. *The Complete Guide to Small-Scale Farming*. Atlantic Publishing Group. Ocala, Fla. 2010. p.197–250.

"New Biopesticide Active Ingredients – 2008." U.S. Environmental Protection Agency. **www.epa.gov/pesticides/biopesticides/product_lists/new_ai_2008.htm**. Accessed on April 23, 2010.

New England Cheesemaking Supply Company. **www.cheesemaking.com**. Accessed on Sept. 6, 2010.

Newman, Y.C. et al. "Creep Grazing for Suckling Calves — A Pasture Management Practice." University of Florida, IFAS Extension. 2009. **http://edis.ifas.ufl.edu/ag193**. Accessed on Sept. 4, 2010.

North American Limousin Foundation. **www.nalf.org**. Accessed on Sept. 5, 2010.

O'Connor, Martin. Chief, Standards, Analysis and Technology Branch. Agricultural Marketing Service, USDA. E-mail on Aug. 31, 2010.

O'Hagan, Maureen. "Slaughterhouse on wheels aids 'locavore' movement." *Seattle Times*. Aug. 11, 2009. **http://seattletimes.nwsource.com/html/localnews/2009636109_slaughterhouse11m.html**. Accessed on May 4, 2010.

Paine, Laura. "Grass-based dairy products: challenges and opportunities." University of Wisconsin–Madison, Center for Integrated Agricultural Systems. August 2009. p.3. **www.kbs.msu.edu/images/stories/Dairy/Grass_Based_Diry_Products_Challenges_and_Opportunities.pdf**. Accessed on June 17.

"PAMTA and STAAR Will Protect Public from Antibiotic Resistance — Letter to Congress." Union of Concerned Scientists and other members of Keep Antibiotics Working Coalition. March 31, 2008. **www.ucsusa.org/food_and_agriculture/solutions/wise_antibiotics/ucs-urges-congress-to-adopt.html**.

Parish, Jane A. "Calving Selection Considerations." Mississippi State University, Extension Service. 2008. **http://msucares.com/pubs/publications/p2501.pdf**. Accessed on March 28, 2010.

Pennington, Jodie A. and VanDevender, Karl. "Heat Stress in Dairy Cattle." University of Arkansas, Division of Agriculture. Cooperative Extension Service Printing Services. p.1. **www.uaex.edu/other_areas/publications/pdf/fsa-3040.pdf**.

Penrose, Christopher D. et al. "Stockpiling Tall Fescue for Winter Grazing." Ohio State University Extension. **http://ohioline.osu.edu/agf-fact/0023.html**. Accessed on Sept. 6, 2010.

Pirelli, G.J. et al. "Beef Production for Small Farms." Oregon State University, Extension and Experiment Station Communications. January 2000. **http://ir.library.oregonstate.edu/xmlui/bitstream/handle/1957/19237/ec1514.pdf?sequence=1**. Accessed on Sept. 6, 2010.

"Plants Database." USDA Natural Resources Conservation Service. **http://plants.usda.gov**. Accessed on April 10, 2010.

Plastic-Mart. **www.plastic-mart.com**. Accessed on Aug. 23, 2010.

Poole, Terry E and Fultz, Stanley W. "Common Plants Poisonous to Livestock in Maryland." University of Maryland, Cooperative Extension Service. p.3. **http://extension.umd.edu/publications/pdfs/fs721.pdf**. Accessed on March 7, 2010.

Pordomingo, Anibal. "Finishing cattle in hot summer areas requires summer annuals." The Stockman Grass Farmer. **www.stockmangrassfarmer.net/cgi-bin/page.cgi?id=426**. Accessed on Sept. 21, 2010.

Powell, Jeremy and Troxel, Tom R. "Stocker Cattle Management: Receiving Health Program." University of Arkansas, Division of Agriculture. **www.uaex.edu/Other_Areas/publications/PDF/FSA-3065.pdf**. Accessed on Oct. 1, 2010.

Pratt, Mindy, and Rasmussen, G. Allen. "Determining Your Stocking Rate." Utah State University, Cooperative Extension. May 2001. **http://extension.usu.edu/files/publications/publication/NR_RM_04.pdf**. Accessed on April 29, 2010.

"Product Labeling." California State University, Chico College of Agriculture and University of California Cooperative Extension. All Things Grass Fed. **www.csuchico.edu/grassfedbeef/regulations/product-labeling.shtml**. Accessed on Sept. 6, 2010.

"Profitable Grazing-Based Dairy Systems." USDA National Resources Conservation Service. May 2007. p.7. **ftp://ftp-fc.sc.egov.usda.gov/GLTI/technical/publications/tn_rp_1_a.pdf**. Accessed on Feb. 16, 2010.

Purebred Dexter Cattle Association of North America. Sept. 21, 2009. **www.purebreddextercattle.org**. Accessed on Aug. 31.

Quaid, Libby. "Critics Have a Beef With USDA's 'Grass-Fed' Plan." *The Associated Press.* Sept. 04, 2006. **http://articles.latimes.com/2006/sep/04/business/fi-grassfed4?pg**. Accessed on Feb. 9, 2010.

"Questions & Answers: Sickness caused by E. coli." Department of Health and Human Services: Centers for Disease Control and Prevention. **www.cdc.gov/ecoli/qa_ecoli_sickness.htm**. Accessed on Sept. 6, 2010.

"Raw Milk Nation." Farm-to-Consumer Legal Defense Fund. **http://farmtoconsumer.org/raw_milk_map.htm**. Accessed on Aug. 30, 2010.

Redfearn, Daren D. and Bidwell, Terrance G. "Stocking Rate: The Key to Successful Livestock Production." Oklahoma Cooperative Extension Service. **http://pods.dasnr.okstate.edu/docushare/dsweb/Get/Document-2050/PSS-2871web.pdf**. Accessed on April 29, 2010.

"Resolution." American Veal Association. May 9, 2007. **www.vealfarm.com/lib/pdf/1225128571.pdf**. Accessed on Sept. 6, 2010.

Rinehart, Lee. "Cattle Production: Consideration for Pasture-Based Beef and Dairy Producers." ATTRA – National Sustainable Agriculture Information Service. p.4. **http://attra.ncat.org/attra-pub/PDF/cattleprod.pdf**. Accessed on Sept. 20, 2010.

Rinehart, Lee. "Ruminant Nutrition for Graziers." ATTRA – National Sustainable Agriculture Information Service. p.16. **www.attra.ncat.org/attra-pub/PDF/ruminant.pdf**. Accessed on Sept. 5, 2010.

Robinson, Jo. *Pasture Perfect.* Vashon Island Press. Vashon, Wash. 2004.

Ruechel, Julius. *Grass-fed Cattle.* Storey Publishing. North Adams, Mass. 2006.

Rumph, J. M. "Understanding EPDs." Montana State University. May 31, 2008. **www.msuextension.org/ruralliving/ls_beef_epd.html**. Accessed on Sept. 6, 2010.

"Rural Development Real Estate for Sale." U.S. Department of Agriculture. **www.resales.usda.gov**. Accessed on Sept. 6, 2010.

Salatin, Joel. *Salad Bar Beef.* Polyface, Inc. Swoope, Va., 1995.

Sechrist; Dan, Peggy, and Richard. "Growing Profit — Factors that Impact Your Bottom Line." Sustainable Agriculture Research and Education. **www.sare.org/publications/beef/factors_bottom_line.shtml**. Accessed on Oct. 1, 2010.

"Selection Tools." Angus Source. American Angus Association. Nov. 3, 2005.

Selk, Glen. "One Calving Season versus Two Calving Seasons." University of Nebraska–Lincoln. Cited by *Beef Magazine*. Feb. 17, 2009. **http://beefmagazine.com/sectors/cow-calf/0218-calving-seasons**. Accessed on May 15.

Soil Test Report for farm in Brandenburg, Ky. University of Kentucky Cooperative Extension Service. Dec. 6, 2007. **http://ces.ca.uky.edu/ces**.

Smajstrla, A.G. et al. "Field Evaluations of Irrigation Systems: Solid Set or Portable Sprinkler Systems." University of Florida, IFAS Extention. **http://edis.ifas.ufl.edu/ae384**. Accessed on Sept. 5, 2010.

Spearman, Becky and Poore, Matt. "Culling the Cow Herd in Drought Situations." North Carolina State University – A&T State University Cooperative Extension. Page 3. **www.ces.ncsu.edu/disaster/drought/cow_herd_culling.pdf**. Accessed on May 13, 2010.

"Spring Compared to Summer Calving." Cooperative Extension System. Feb. 18, 2008. **www.extension.org/pages/Spring_Compared_to_Summer_Calving**. Accessed on Aug. 31, 2010.

Stoltzfoos, Dennis. Personal interview on March 26, 2010.

Thom, W.O. et al. "Taking Soil Test Samples." University of Kentucky. December 2003. **http://www.ca.uky.edu/agc/pubs/agr/agr16/agr16.pdf**.

Umphrey, J. E. and Staples, C.R. "General Anatomy of the Ruminant Digestive System." University of Florida, IFAS Extension. **http://edis.ifas.ufl.edu/ds061**. Accessed on Feb. 26, 2010.

"United States Standards for Grades of Carcass Beef." Agricultural Marketing Service. January 31, 1997. U.S. Department of Agriculture._ **www.ams.usda.gov/AMSv1.0/getfile?dDocName=STELDEV3002979**. Accessed on June 17, 2010.

"United States Standards for Livestock and Meat Marketing Claims." U.S. Department of Agriculture. Federal Register, Vol. 67, No. 250. December 30, 2002. **www.ams.usda.gov/AMSv1.0/getfile?dDocName=STELDEV3102210.** Accessed on April 2.

"United States Standards for Livestock and Meat Marketing Claims, Grass (Forage) Fed Claim for Ruminant Livestock and the Meat Products Derived From Such Livestock." U.S. Department of Agriculture. Federal Register, Vol. 72, No. 199. Oct. 16, 2007. **www.ams.usda.gov/AMSv1.0/getfile?dDocName =STELPRDC5063842**. Accessed on Aug. 30, 2010.

"U.S. Climate Normals." National Climactic Data Center. **http://cdo.ncdc. noaa.gov/cgi-bin/climatenormals/climatenormals.pl**. Accessed on April 10, 2010.

U.S. Small Business Administration. **www.sba.gov**. Accessed on Oct. 1, 2010.

Voth, Kathy. "Cows Eat Weeds!" Livestock for Landscapes and Utah State University. **www.cfc.umt.edu/cesu/NEWCESU/Assets/Newsletters/2009/ COWS%20EAT%20WEEDS.pdf**. Accessed on Sept. 4, 2010.

Vough, Lester R. "Evaluating Hay Quality." University of Maryland Cooperative Extension. **http://extension.umd.edu/publications/pdfs/fs644.pdf.** Accessed on Sept. 6, 2010.

"Watering Systems for Serious Graziers." Missouri Natural Resources Conservation Service. 2006. **http://www.mo.nrcs.usda.gov/news/pubs_ download/out/Watering%20Systemslow.pdf**.

Weary, Dan. "Reducing Pain Due to Caustic Paste Dehorning." **http:// farmwest.com/index.cfm?method=pages.showPage&pageid=457**. Accessed on June 15, 2010.

White, Rob. Personal interview on March 21, 2010.

"Why Ayrshires." U.S. Ayrshire Breeders' Association. **http://www.usayrshire. com/whyayrshire.html**. Accessed on Sept. 5, 2010.

"Why Galloway?" American Galloway Breeders Association. **http:// americangalloway.com/why.php**. Accessed on Sept. 5, 2010.

Wiedmeier, Randall D. et al. "Cow-Calf Production and Profitability on Irrigation Pastures Composed of Forage Mixtures or Monocultures During the First Year After Establishment." Utah State University Extension. May 2004. **www.behave.net/BFN/Research%20updates/randy-past-wt-gain.pdf**. Accessed on Aug. 31, 2010.

Wisconsin Grass-fed Beef Cooperative. **www.wisconsingrassfed.coop**. Accessed on Sept. 21, 2010.

Wood, John. Personal interview on Feb. 24, 2010.

Wright, C.L. and Pruitt, R.J. "Fenceline Weaning for Beef Cattle." South Dakota State University College of Agriculture and Biological Sciences. October 2005. **http://pubstorage.sdstate.edu/AgBio_Publications/articles/ExEx2049.pdf**. Accessed on Aug. 27, 2010.

Yegerlehner, Alan. Personal interview on April 13, 2010.

"Your Questions Answered..." The Stockman Grass Farmer. September 12, 2006. **www.stockmangrassfarmer.net/cgi-bin/page.cgi?id=367**. Accessed on Feb. 22, 2010.

Index

A

B

C

D

E

F

G

H

R

S

U

V

W

Y